Technical, Economic and Societal Effects of Manufacturing 4.0

Mikael Collan • Karl-Erik Michelsen
Editors

Technical, Economic and Societal Effects of Manufacturing 4.0

Automation, Adaption and Manufacturing in Finland and Beyond

palgrave
macmillan

Editors
Mikael Collan
School of Business and Management
LUT University
Lappeenranta, Finland

Karl-Erik Michelsen
School of Business and Management
LUT University
Lappeenranta, Finland

ISBN 978-3-030-46102-7 ISBN 978-3-030-46103-4 (eBook)
https://doi.org/10.1007/978-3-030-46103-4

This Palgrave Macmillan imprint is published by the registered company Springer Nature Switzerland AG.
The registered company address is: Gewerbestrasse 11, 6330 Cham, Switzerland

Preface

This book is a result of cross-disciplinary teamwork around a common theme, Manufacturing 4.0 and the technical, economic, and social issues that go with it. In vein with the original idea of Industry 4.0, put forth as a vision of a new kind of future for the industry that encompasses also the many societal and economic changes that come with technological change, this book studies the technological change in manufacturing and its possible transformative power on the society.

Most importantly the message in the book is that Manufacturing 4.0 is not only a technical change, nor is it a purely technically driven change, but it is a societal change that has the potential to disrupt the way societies are constructed in both in the positive and in the negative.

We hope this book both reinforces the understanding of the phenomenon that is Manufacturing 4.0 and opens the readers′ eyes on what more it could be.

Lappeenranta, Finland

Mikael Collan
Karl-Erik Michelsen

Acknowledgements

The editors wish to acknowledge the help received the creation of the book by numerous individuals in the industry and in academia. Special thanks for their constructive input go to Dr. Paolo Bosetti, Simone Muraglia, and professor Dr. (Med. & Chir.) Seppo Junnila and to Anna Huusko for her support with the editorial tasks.

The support of the Finnish Strategic Research Council project "Manufacturing 4.0" is gratefully acknowledged—without the support the creation of this book would not have been possible.

Contents

Notes on Contributors

Henri Aaltonen works in the Finnish Strategic Research council project Manufacturing 4.0 at the University of Helsinki. His background is in European studies, international relations, and in international business. Aaltonen's research has focused on European integration, especially from the point of view of the Finnish paper industry.

Matteo Brunelli obtained his Bachelor and Master degrees from the University of Trento, Italy, and his doctoral degree from Åbo Akademi University, Finland. He spent five years as a Post-doctoral researcher at Aalto University, Finland, and is presently an associate professor at the department of industrial engineering of the University of Trento, Italy. His research interests include decision analysis, preference modeling, mathematical representations of uncertainty, and applied optimization.

Markus Chmielus is an Associate Professor in the Mechanical Engineering and Materials Science Department since 2013 with a background in Materials Science and Engineering (M.Sc. from Boise State University, ID, USA; PhD from the Technical University Berlin/Helmholtz Center Berlin for Materials and Energy, Germany; post-doctoral researcher at Cornell University, NY, USA) and Aerospace Engineering (Dipl.-Ing. from the University of Stuttgart). Dr. Chmielus's current research includes investigating the influence of production and processing parameters on the microstructure and properties of metals, carbides and functional magnetic materials produced via additive manufacturing (3D printing), as well as via thin film deposition. The combining umbrella of all his research is quantitative, correlative characterization of microstructure, defects, and properties on different length-scales. His work on functional magnetic materials including magnetic shape memory alloys and magnetocaloric materials began in 2004 with single crystals and foams and expanded to beam and non-beam additively manufactured materials.

Mikael Collan is a tenured professor of strategic finance at Lappeenranta-Lahti University of Technology in Lappeenranta, Finland. He has a D.Sc. in information systems from Åbo Akademi University, Finland. He is the academic director of the master´s programme in strategic finance and business analytics and the founder and co-director of two degree programmes in business analytics, one in engineering and one in business. He is an ordinary member of the Finnish Society of Sciences and Letters, the oldest of the four Finnish academies of Science and the past president of the Finnish operations research society. His research concentrates on business model analysis, investment decision-making, application of analytics in business and in healthcare, and the application of fuzzy logic in business modeling. He has written over fifty journal articles, twenty book chapters, and all in all about two hundred academic publications. He serves on the boards of numerous SME companies.

Raija Hämäläinen is a professor at the department of Education at the University of Jyväskylä, Finland. She works in the field of technology-enhanced learning at the Center for Research for Learning and Teaching. Her research interests include collaboration and creativity at the technology-enhanced learning settings, workplace-learning, and teacher-student interaction. She has a number of international collaborations and serves as an associate editor of Educational Research Review.

Heikki Handroos is a professor of machine automation at the LUT University, Lappeenranta, Finland. Since 1993 he heads the Laboratory of Intelligent Machines at LUT University. His research interests range from modeling, simulation, and control of mechatronic systems to robotics, hybrid transmission, and mobile machinery. He has published about 240 scientific journal- and conference papers in the field of mechatronics. He has been a visiting Professor in University of Minnesota, National Defense Academy (Japan), and at the St. Petersburg Peter the Great Polytechnic University in Russia. He is a member of ASME and IEEE.

Heikki Hiilamo works as a professor of social policy at the University of Helsinki and as a research professor at the National Institute for Health and Welfare. Previously Hiilamo has worked as research professor at the Social Insurance Institution on Finland and as visiting professor in UC in San Francisco and at VID in Oslo, Norway. Hiilamo has the title of Docent from the University of Tampere and the University of Eastern Finland. His research interests include family policy, poverty, inequality, welfare-state research, and tobacco control.
In 2016 Hiilamo was awarded with State Award for Public Information.

Jari Kaivo-oja is Research Director at the Finland Futures Research Centre at the University of Turku. He is an Adjunct Professor at the Faculty of Science at the University of Helsinki (planning and management sciences of regional planning) and the Department of Social Sciences at the University of Lapland (foresight and innovation research). Dr. Kaivo-oja is a Research Professor Kazimieras Simonavicius University in the Platforms of Big Data Foresight project, Vilnius, Lithuania and a researcher at the Strategic Research Council (SRC) of the Academy of Finland in Manufacturing 4.0 and EL-TRAN projects. He has worked in various foresight and innovation management projects for the European Commission, for the European Foundation, for the Nordic Innovation Centre and for the European Parliament.

Mikkel Stein Knudsen is a Project Researcher at the Finland Futures Research Centre at the University of Turku. He has a background in Political Science, and he has a long history of working with private company businesses aiding transformations and competitiveness before joining the field of futures research. In addition to the Manufacturing 4.0-project, he has acted as researcher and as country manager in various international research projects.

Markus Korpela is a doctoral student at LUT University's mechanical engineering department and working as a Junior Researcher at research group of Laser Material Processing & Additive Manufacturing. Mr. Korpela's doctoral studies and research focus on laser-based powder bed fusion from the perspectives of metal material properties.

Kalevi Kyläheiko is Professor of Economics (Emeritus) at Lappeenranta University of Technology, School of Business, Finland. His research interests cover, among others, economics-based strategy research, internationalization of firms, and the use of real options in the strategic management context. He has published about 80 articles in refereed journals and books. Professor Kyläheiko is also an Honorary Professor at the University of Bremen (Emeritus), Germany.

Ville Laitinen is a junior researcher and a second-year doctoral student in Material Physics Laboratory at Lappeenranta-Lahti University of Technology (LUT), Finland. He obtained his Master's degree of Mechanical Engineering in 2015 at LUT specializing in additive manufacturing and laser materials processing. His major research activities are related to laser additive manufacturing of Ni-Mn-Ga based magnetic shape memory alloys (MSMAs). His other research interests include development of MSMA applications, welding of additively manufactured structures and use of pulsed lasers in powder bed fusion based processes. Professionally, he has worked as an automotive industry intern at XOR Motors SAS in Valbonne, France during 2015–2016. He has more than 5 years' experience in additive manufacturing related activities, including consulting and steering group memberships in two regional third party projects incorporating additive manufacturing. He is a member of Finnish Rapid Prototyping Association and The Minerals, Metals & Materials Society.

Theresa Lauraeus is a part-time senior researcher at the Finland Futures Research Center at the Turku University School of Economics and a part-time Wihuri Foundation post-doc researcher at the Department Information and Service Management, Aalto University School of Business. She is also a part-time Chief Researcher in Kazimieras Simonavicius University, Vilnius, Lithuania. Her current research concentrates on technological innovations and future and foresight research.

Ming Li is a post-doctoral researcher at the LUT University in Lappeenranta, Finland. Since 2015 he has been a research-engineer connected to the EUROfusion Engineering Grants (EEG) under the implementation of Fusion Roadmap in Horizon 2020. His research interests cover robotics, artificial intelligence, control system design, embedded systems, and software development. He was a visiting researcher in RACE of UKAEA (UK Atomic Energy Authority) in 2016.

Päivi Maijanen is Post-Doctoral Researcher at the School of Business and Management at the Lappeenranta-Lahti University of Technology in Finland. Her research interests are focused on strategic management and organizational change, and in particular, on dynamic capabilities in the context of digital transformation. She has published in strategy research journals. She has been a visiting scholar via SCANCOR (The Scandinavian Consortium for Organizational Research) at Stanford University, in Jönköping International Business School, and at Tallinn University.

Mahdi Merabtene is a second-year Master's degree student in Welded Metal Structure Programme, Mechanical Engineering at Lappeenranta-Lahti University of Technology (LUT). His research interests include laser processing, additive manufacturing, materials engineering and district cooling and heating energy systems. Professionally, he has worked at DC Pro Engineering in Dubai as mechanical design engineer in several district cooling megaprojects including Qatar World Cup Stadiums, KAUST University, Dubai Creek and many others.

Karl-Erik Michelsen is professor at LUT University specializing in innovation and technology studies and business history. He has worked in numerous national and international projects funded by the Finnish Academy and The EU. He has studied global and international companies and how family companies become global corporations. He has just finished an international project (Horizon 2020) that focuses on the interaction between nuclear industry and the civil society. In 2018 Michelsen was selected the "professor of the year" in Finland. He is a member of several academic societies.

Andrija Milojević is a post-doctoral researcher at the Laboratory of Intelligent Machines at the LUT University, where he has worked since 2017. Previously he has worked as a researcher at the Faculty of Mechanical Engineering, University of Nis, Serbia. His main fields of research are mechatronics and robotics.

Hanna Nygren is a project researcher and a doctoral candidate at the Finnish Institute for Educational Research at the University of Jyväskylä, Finland. Her research focuses on VET-educated adults′ adaptation to technology. Previously she has worked as a drama- and teacher educator.

Olli Nyrhilä has been working with additive manufacturing since beginning of 90s in EOS Finland Oy, Turku, Finland. Nowadays he is responsible leader of R&D of metal material development in EOS GmbH. EOS

Finland has c. 70 employees and c. 20 metal AM machines in Turku, Finland.

Dario Petri is a full professor of instrumentation and measurement and the head of the department of industrial engineering at the University of Trento, Trento, Italy. He is an IEEE Fellow and the recipient of the 2020 IEEE Joseph F. Keithley Award "for contributions to measurement fundamentals and signal processing techniques in instrumentation and measurement." Dario Petri served as the president of the Italian scientific Association of Electrical and Electronic Measurement from 2013 to 2016 and the IEEE Italy Section from 2012 to 2014. He is also the chair of the IEEE Smart Cities Initiative in Trento since 2015. He is the author of over 350 scientific papers published in international journals or in proceedings of peer reviewed international conferences. His research activities are focused on digital signal processing, data acquisition systems, and fundamentals of measurement theory.

Heidi Piili is working as a professor at the Research Group of Laser Materials Processing (LUT Laser) at LUT University, Lappeenranta, Finland. Piili has been working more than 18 years with laser processing of various materials and additive manufacturing, mostly concentrating to powder bed fusion of metallic materials. Her doctoral thesis in 2013 was about characterization and mathematical modeling of interaction between laser beam and material. Piili was working within industrial additive manufacturing during year 2017 at Electro Optical Systems (EOS) Finland, and her main task was to develop monitoring of additive manufacturing of metallic materials. At the moment, she participates to national and international projects (academic and industrial) and also educates additive manufacturing and laser technology. She has more than 115 international publications in field of laser technology and additive manufacturing.

Juhani Rautopuro is the head of the research-area of "educational systems and society-research" at the Finnish Institute for Educational Research at the University of Jyväskylä, Finland. His main focus is on the international assessment studies, including PISA, TIMSS, and PIRLS in Finland and the main research areas are assessing learning outcomes in general education, equitable student assessment, teaching and learning of quantitative methods in education and the risk of exclusion of education. He is also an adjunct professor on evaluation of education at the University of Helsinki. Previously he has worked at the Finnish National Board of Education and at the Finnish Education Evaluation Centre.

Niko Riikonen is working as Project Researcher at research group of Laser Materials Processing & Additive Manufacturing in LUT University. Riikonen has worked in the field of laser microprocessing including laser marking and laser engraving of metal materials. Currently he is working in the field of laser-based powder bed fusion of metal materials.

Antti Salminen is a professor at the Department of Mechanical Engineering, University of Turku, Turku, Finland and has more than 30 years' experience in laser processing of various materials. He has been acting as researcher and project leader in several academic studies and working with industrial cases, in the field of process development, laser system development, product design for laser processing, consulting and technology transfer. He has published more than 400 papers and of those over 300 scientific. Currently he is running projects for digitalization of production and additive manufacturing. He is member of the board of Finnish Welding Association and Finnish Additive Manufacturing Association FIRPA.

Jyrki Savolainen is as a post-doctoral researcher at the School of Business and Management at the LUT-University, Lappeenranta, Finland. He has a D.Sc. (Econ. & Bus. Adm.) from the LUT-University and a M.Sc. (Tech.) from the University of Oulu, Finland. His research interests include business model analysis and the application of system dynamic simulation in industrial problems. Specific fields of interest include metals mining, additive manufacturing, and the mechanical forest industry.

Erica Stevens is a fifth-year PhD student in Materials Science and Engineering at the University of Pittsburgh in Pittsburgh, Pennsylvania, funded by the Army Research Laboratory through the National Defense Science & Engineering Graduate Fellowship. Her research experience is primarily in binder jet additive manufacturing, particularly of functional magnetocaloric Ni-Mn-based materials, and the microstructural and magnetic characterization thereof. Stevens earned her Bachelor's degree in Materials Science and Engineering from the University of Pittsburgh in 2015, completed an internship at Pacific Northwest National Laboratory in 2015, and was a visiting researcher at the University of the Basque Country (EHU/UPV) in Bilbao, Spain during 2018. Stevens has been first author or co-author on over 20 publications, and has given 18 talks and poster presentations. She has placed in four poster contests, has won numerous scholarships, and was elected as President of the Microscopy Society of America Student Council for 2019–2020.

Kari Ullakko is Professor and Head of the Material Physics Laboratory at Lappeenranta-Lahti University of Technology, Finland. He has worked with several universities and companies in various countries and has led a number of international research teams and projects (Finland, EU and USA). His discovery of the magnetic shape memory effect has created a new field of material science, engineering and technology. Ullakko's first paper in this field has been cited over 2700 times. Dr. Ullakko is an author of over 300 publications and an inventor in 12 patents. Ullakko has founded two high technology companies. Ullakko is a member of several international professional organizations and advisory committees.

Michele Urbani is a junior researcher at the Department of Industrial Engineering at the University of Trento in Italy. He has a M.Sc. (Eng.) from the University of Trento. He is working on his dissertation within the broad field of maintenance modeling with a focus on how analytics and additive manufacturing may affect the field maintenance.

Jan Van Humbeeck obtained his Master's degree of Materials Engineering in 1976 and his PhD in 1983 at KU Leuven (Belgium). Since 2017, he is prof. emeritus at KU Leuven. His main research topic was related to the development of optimization of structural and functional properties of Shape Memory Alloys (Cu- Fe- NiTi-based). Since about 15 years, he devotes a large part of his time also to microstructural analysis and optimization of the properties of parts produced by Additive Manufacturing. The focus is on SLM (Selective Laser Melting) of specific Titanium-, Aluminum-, Steel and NiTi-alloys. Jan Van Humbeeck published about 450 papers in peer-reviewed journals, about 200 papers in conference proceedings and 14 chapters in books, and gave more than 100 invited and keynote lectures and seminars. He was visiting professor at research institutes and universities in Argentina, India, Japan, Spain and Switzerland.

Maarit Virolainen is a post-doctoral research at the Finnish Institute for Educational Research at the University of Jyväskylä, Finland. The focus of her research is on vocational and professional education, learning from work experience, and on post-compulsory education learning pathways—international comparison of VET systems and their development are of particular interest. She is a member of board of NORDYRK (Nordic research network on vocational education and training) and has served as the head of the Finnish Vocational Educational Research Association. Recently she has participated in the comparison of vocational education and training models in the Nordic countries in terms of how they enable transitions to the world of work and higher education.

List of Figures

List of Tables

Industry 4.0 in Retrospect and in Context

Karl-Erik Michelsen

1 Coming to Terms with the Concept of Industrial Revolution

What is an industrial revolution? Does the concept take us back to the history or does it carry us to the future of industrial manufacturing? Quick Google search gives us both perspectives. Industrial revolution connects the past, the present, and the future.

It is widely agreed that the Internet, artificial intelligence, Internet of things, automated robots, sensors, augmented reality, Big Data and several other groundbreaking innovations will configure global industrial landscape. Industrial companies can collect, analyze and process data and use the Internet and advanced ICT for the manufacturing of high quality industrial goods.

Current changes in manufacturing systems are significant, but not exceptional in history. Many radical and even more frequent incremental changes in manufacturing systems have taken place during the two centuries that sophisticated machines have been used in manufacturing processes. Sometimes radical changes disrupt the evolutionary path and the

K.-E. Michelsen (✉)
School of Business and Management, LUT University, Lappeenranta, Finland
e-mail: karl-erik.michelsen@lut.fi

M. Collan, K.-E. Michelsen (eds.), *Technical, Economic and Societal Effects of Manufacturing 4.0*,
https://doi.org/10.1007/978-3-030-46103-4_1

reigning manufacturing paradigm breaks down—a new type of manufacturing system is established and the new era is typically coined as another "Industrial Revolution". Although several paradigm changes have taken place since the late eighteenth century, it is still unknown, what mechanisms drive these changes. As Peter Temin (1997) and other scholars have claimed, the term "industrial revolution" is itself either too vague to be of any use at all, or it produces false connotations of abrupt changes [1].

Although the mechanisms of change that drive industrial development are still unknown, the concept of industrial revolution is widely used in popular literature, textbooks, and in policy documents. The human mind likes to bring structure in to the chaotic past and the evolution of industrial manufacturing is commonly divided in three or four chapters:

The First Industrial Revolution started in Britain during the latter part of the eighteenth century. The steam engines were invented and connected to textile looms. As a result, the manufacturing of consumer goods changed from the individual and domesticated setting into the factory. Steam engines were gradually applied to other sectors of production, then to transportation, and finally to production of energy. The rate of change was slow and it took decades before the new manufacturing system resulted in radical changes in the society. The first industrial revolution was a local phenomenon that spread from the Great Britain to Western Europe, and to the United States.

The Second Industrial Revolution started approximately hundred years later in the United States. The discoveries of electricity, the combusting engine, the telephone, and innovations in chemical and material technologies sparked "The Great Leap Forward". A technological torrent flushed over industrialized nations and the manufacturing systems expanded rapidly, both vertically and horizontally. Large scale factories were built and connected via transmission lines to the centralized power and heat producing plants and started exchanging information through telecommunication networks. Henry Ford developed a non-stop manufacturing system and applied it to his automobile factories in Detroit. Mass production system of manufacturing changed the division of labor within factory walls. Semi-skilled workers operated along the assembly line, which poured out highly standardized industrial goods. The Second Industrial Revolution spread across the Atlantic Ocean, developed further in Western Europe and Scandinavia, and entered into the Soviet Union.

Less than hundred years later, traditional industries based on oil and fossil-fuels and on mass production could no longer be considered a complete solution to the economic or the social problems. A cluster of radical innovations in communication- and energy-technologies merged into a new economic era. A powerful new infrastructure was created for manufacturing industries. What is the *Third Industrial Revolution* is still an ongoing process, but visible changes are already taking place in advanced industrial economies. Centralized production systems and energy production networks are challenged by de-centralized systems that engage hundreds of millions of people to produce their own green energy in their homes, offices, and factories and to share information online [2].

Although the Third Industrial Revolution is still in its infancy, the *Fourth Industrial Revolution* is already knocking at the door. Advanced ICT applications, Big Data, industrial robotics, and automated production systems will merge into Cyber-Physical (CPS) systems, which provide new platforms and infrastructure for manufacturing industries. The concept of Industry 4.0, or the Fourth Industrial Revolution, is recognized in national and international forums and the change-process is included in the major policy documents and development programs [3].

Although, it is widely recognized that the classification of industrial development into three or four "industrial revolutions" is inadequate and even misleading, the concept is still used to demonstrate, how the manufacturing industry has changed over time. The concept of "Industrial Revolution" is also used for propaganda purposes to promote technological enthusiasm and radical changes in the industrial landscape.

What drives industrial change? This question is widely debated among economists, historians, and social scientists. For engineers, the problem is less difficult to answer. From the engineering point of view, the evolution of industrial production can be viewed through the lens of technology. Technological change drives social changes and the accumulation of radical innovations cause disruptions in manufacturing systems. This type of argumentation defines technology as an autonomous phenomenon in society—technology accumulates according to the deterministic laws, which are dictated by scientific knowledge.

Classification of manufacturing systems is shaped by technological determinism. As Boyd and Holton [4] conclude, the changes in manufacturing systems are generated by the changes in technology. Technological determinism narrows down the argumentation into technical details and technological solutions. Technological determinism bypasses social issues, which are hidden conditions of existence of and for the new technologies.

Technological determinism is challenged by social constructivists, who argue that technology is (also) shaped by social forces, ideologies, and values. Although technology has the ability to change social structures and human behavior, the effect is never a one-way-street. Same technological applications are used and shaped differently depending on social, cultural and political environments. According to David Noble [5] manufacturing systems are more than technological artifacts. They are social processes that reflect social structures, values, and ideologies. Manufacturing systems are designed to shape societies, change power structures, and to benefit particular economic and ideological goals. The interaction between technology and society is a systemic process, where social forces shape the systems and the systems shape social forces.

Hence, in order to understand what is meant by Industry 4.0, or the Fourth Industrial Revolution, the phenomenon must be placed in the historical context. Manufacturing systems are complex large technological systems, which are managed by highly trained professionals. As Thomas Hughes [6] points out, large manufacturing systems contain messy, complex, problem-solving components, which are physical artifacts, but they also contain organizations and immaterial components. Manufacturing systems are socially constructed and adapted in the society in order to function effectively. They use natural resources, but also social, political and cultural resources, such as knowledge, legislation, regulation, and ideology. Manufacturing systems contribute to the development of modern industrial societies, but while doing this, they become depended and intimate parts of modern industrial and post-industrial societies.

2 Unde Venis Industry 4.0?

"Industry 4.0" was introduced at a press conference at the Hannover Fair, in 2011. Three German engineers, Henning Kagermann (SAP), Wolfgan Wahlster (Professor of artificial intelligence), and Wolf-Dieter Lucas (Senior officer at the German Ministry of Education and Research) introduced a vision of the future manufacturing system, "The Industry 4.0". The idea was received with enthusiasm in Germany and the concept spread rapidly to other European countries. Five years later, it had already gained international recognition and The World Economic Forum in Davos organized a thematic session titled "Mastering the Fourth Industrial Revolution".

The first strategic paper drafted by Henning Kagermann and others [7] defined the goals of Industry 4.0 as follows: "*It will address and* solve some of the challenges facing the world today *such as resource and energy efficiency, urban production, and demographic change. It enables continuous* resource productivity and efficiency *gains to be delivered across the entire value network. It allows work to be organized in a way that takes demographic change* and *social factors into account. Smart assistance systems release workers from having to perform routine tasks, enabling them to focus on creative, value-added activities. In view of the impending shortage of skilled workers, this will allow older workers to extend their working lives and remain productive for longer. Flexible work organization will enable workers to combine their work, private lives, and continuing professional development more effectively, promoting a better* work-life balance".

As Kagermann and others' [7] strategic paper demonstrated, Industry 4.0 is not only a technological platform for manufacturing industry. It is a social program, which targets major social, economic, and political challenges in the twenty-first century. The concept reflects also political and economic discourses, which have taken place in Germany, Europe, and North America since the international financial crises in 2008. The collapse of the global financial systems drove Western countries into an economic and political chaos, which had long lasting consequences. Global economy, free and almost unregulated flows of goods, capital and knowledge divided the world into winners and losers.

Western industrial countries had gone through a rapid de-industrialization process, when large swathes of manufacturing industries moved to China, India, and to other developing countries. The transformation was regarded as a positive and a progressive turn in the development ofWestern economies, which focused on the information technology, services, innovations, and on the creation of useful knowledge. Mass production of industrial goods was no longer viewed as a necessary part of the economy. Instead, the 'chimney industries' polluted the environment and offered monotonous and unattractive jobs for low salary workers. However, as Joseph Heathcott and Jefferson Cowie [8] have concluded, de-industrialization was a much broader and fundamental transformation than anybody had anticipated. It turned out to be a socially complicated, historically deep, geographically diverse, and a politically perplexing phenomenon [9].

When factories were closed and abandoned in the Western world, new production facilities and manufacturing systems were rapidly erected in

China. The growth of industrial production in China alone was spectacular and in less than 40 years, China had become the factory of the world. Chinese factories produced about 50% of the world's major industrial goods. This reflected to the GDP, which grew more than 10% annually.

Meanwhile, Western economies struggled to cope with the unprecedented consequences of globalization and de-industrialization. After financial crises, the unemployment rates stayed high, economic growth was slow and deficits climbed to alarmingly high numbers. The political climate that had hailed the destruction of walls and barriers, open borders, and economic liberalization was suddenly challenged by neo-nationalism which tried to put a stop to global flows and restore the nation-state as the sovereign political and economic actor in society [10].

Reflecting the robust economic growth in China and in other Asian countries, the European Union and its member states tried to find ways to restructure the economic landscape. It was understood that the post-industrial information society did not alone provide a stable base for economic growth. Emerging industrial economies in Asia threatened to take over global markets and with the accumulation of capital to develop the next generation of technologies and industry, which would undermine the competitiveness of European corporations. These concerns were reflected in the ambitious research and development projects, which were initiated, organized and funded by the European Union under the Horizon 2020 umbrella. Europe was looking for a new momentum that would reverse the uneven growth and sluggish job recovery that resulted from the financial crises. The long term structural change called for an industrial recovery that would be based on smart, sustainable and technologically advanced manufacturing systems [11].

The strategy paper written by the advocates of the Industry 4.0 answered to this call. The attractive idea did not die after the Davos conference, but instead it spread out to the business community in Germany. The concept of Industry 4.0 and The Fourth Industrial Revolution was appealing to the managers and CEOs who sought to find ways to improve productivity and to increase production of high technology goods. Soon after, the political actors at the European Union and at the national level got engaged in the process and pushed it forward to the policy programs. Large corporations sensed the opportunity and adopted the narrative of the Industry 4.0 and the Fourth Industrial Revolution into their own strategy papers [12].

Hence, the basis for the concepts of Industry 4.0 is not a concrete and realistic analysis of the transformation that has taken place in the manufacturing industry—but instead, the Industry 4.0 is an effort to control the future, which is full of uncertainties and discontinuities [12]. The concept is not a concrete platform, but rather a vision used in the future-making process. Industry 4.0 tells the audience, how the future industries are organized to fulfill global strategic goals. This takes place in the globally connected and almost autonomously functioning manufacturing units. This future vision is competitive and it will overpower traditional mass production systems, which are neither ecologically nor socially sustainable. When the Industry 4.0 is applied globally, it will tame the future by reorganizing the production of industrial goods with new roles of and for human labor.

By now, Industry 4.0 has gained momentum beyond Germany. United States government has organized a series of discussions on the Advanced Manufacturing Partnership (AMP). German government passed in 2012 the High-Tech Strategy that targeted billions of euro to the development of cutting-edge technologies. The following year the French Government initiated "La Nouvelle France Industrielle" program and the UK Government followed the lead with the long-term program "Future of Manufacturing". The European Commission launched the new contractual Public-Private Partnership on "Factories of the Future" program in 2014. South Korean government joined the process in 2014 by announcing "Innovation in Manufacturing 3.0" program. The Chinese government followed the year after with the "Made in China 2025" strategy and the "Internet Plus" programs. The Japanese Government adopted the "5th Science and Technology Basic Plan", which included the "Super Smart Society" program. The following year the Singapore government invested $19 billion to the Research, Innovation, and Enterprise (RIE) plan. In addition, the American high technology corporations AT&T, Cisco, General Electric, IBM, and Intel established "The Industrial Internet Consortium". Similar collaboration was established between German, Japanese, and American high technology companies.

When we look back in the history of industrial manufacturing, the concept of Industry 4.0 does not fit in the typical pattern. It took more than hundred years before the term "First Industrial Revolution" was coined and the Second Industrial Revolution was defined approximately half a century after the take-off. Both revolutions had matured and gained momentum and both negative and positive consequences of the transformation could be observed and analyzed. The economic consequences of the first industrial revolution were significant, but it took several decades before changes in productivity and the GDP could be measured. Social consequences of the first industrial revolution were harsh and they were thoroughly analyzed by Karl Marx and Friedrich Engels among others already in the middle of the nineteenth century.

The second industrial revolution was monitored closer from the very beginning by the modern news media, which brought the radical innovations and heroic innovators to the international limelight. Economists and social scientists documented changes in national economies and social conditions. Future scenarios were painted, and for the first time, industrial production, manufacturing systems, and supporting infrastructure were included in the policy documents. Positive effects of the Second Industrial Revolution were contrasted to the negative effects of mass production, centralization of production and distribution, and the massive consumption of natural resources. These two contradictory pictures were embodied in the modern industrial society, which replaced slowly and gradually the traditional agricultural societies.

If compared to the first two industrial revolutions, the situation today is very different. Third and Fourth Industrial Revolutions provide very little concrete evidence for the economic and social analyses. Both concepts promise radical changes in the manufacturing systems and major changes in the organization of work and of everyday life. The positive scenarios promise rapid increases in productivity, sustainable production, and higher standards of living. Negative scenarios predict massive elimination of work and many of the current professions, which have formed the foundation of modern societies. How fast these changes will take place? If we draw conclusions from the past experience, the change will be slow and gradual. As Brynjolfsson and others [13] have demonstrated, transformation from one technological system into another was long delayed and far from automatic business.

In fact, we know very little about Industry 4.0 and the Fourth Industrial Revolution. What we know is more about the vision for the future than about actual analysis of the current situation. Conclusions about the future are drawn from the past experience and for this purpose the "Industrial Revolution" is a credible concept. However, it is worth remembering that originally the concept was used, when trying to understand the multiple consequences of industrial production. Today, the advocates of the Industry 4.0 and the Fourth Industrial Revolution try to harness the future, which is unpredictable and chaotic. If the concept of Industry 4.0 is placed in this context, it becomes more of a political and an ideological concept, than a blueprint of the technological future.

3 Implications of Industry 4.0 Beyond Technology

What are the social implications of the Industry 4.0.? If this question is viewed from the historical point of view, the answer is less than certain. Industrial revolutions from the late eighteenth century to today have caused massive and mostly unpredictable social changes. Industrial production is based on the interaction between machines and humans, which takes place in the factory. It is a carefully designed space for productive manufacturing of industrial goods. Factory is an artificial environment, which reflects the functions of machines. Manufacturing systems, on the other hand, are both abstract and concrete blueprints, which describe how the flows of raw materials and energy are turned into products. Human labor collaborates with the work of the machines. Machines are able to work without breaks almost 24/7, but human labor must be scheduled differently. How this is done, and who has the power to decide about the division of labor within the factory walls has been, and still is, one of the most heated political, social, and ideological issues [14].

The interplay between machines and human labor started slowly, but escalated during the Second Industrial Revolution. According to von Tunzelman [15], the Second Industrial Revolution integrated useful scientific knowledge into technological developments and brought the results from the collaboration into the factory. Radical innovations in energy technology allowed long distance transfer of electricity to factories. Consumption traditions changed and standardized and inexpensive industrial goods replaced uniquely crafted hand-made products. Factories were organized to follow the philosophies of economies of scale and throughput. The best example of this was the mass production system, which is based on the American System of Manufacturing. Continuous moving belts moved interchangeable parts on the belt, where they were assembled into standardized industrial goods. This production system integrated the human work and the machine work into a seamless web. Production systems required a great deal of coordination and understanding on how human physiology and psychology could be optimized to serve the manufacturing system.

The use of human labor in the manufacturing system has been a permanent challenge since the beginning of the industrial era. Human skills are important and skilled workers are needed to supervise the production and to manage complicated issues. However, human labor is the irrational part of the manufacturing system. Every worker is different and the capacity of workers changes from day to day and from year to year. On the other hand, machine work can be standardized and if managed properly, machines operate without disruptions. Hence, the manufacturing systems have tried to minimize and even eliminate human labor from the system. Fredrik W. Taylor introduced scientific management into factories in order to standardize work and to find the "right man for the right work". During latter part of the twentieth century, automated machines and robots started to take over human work. Information technology and advanced ICT applications have escalated this development and many factories are currently operating almost without human assistance [16].

The Industry 4.0 promises to make even more radical changes in the human-machine collaboration. It is not so far in the past, when the automobile factories employed more human workers than machines. Now the balance has shifted and less and less human labor is needed in the shop-floor level. The technologies of Industry 4.0 have the potential to erase the labor issue by substituting human workforce with an army of robots, automated systems, and algorithms. This applies equally to the semi-skilled and qualified workers, who must compete against intelligent machines and smart systems.

Industry 4.0 will also change the future of professional work. According to Richard and Daniel Susskind [17]: "*In relation to our current professions, we argue that the professions will undergo two parallel sets of changes. The first will be dominated by automation. Traditional ways of working will be streamlined and optimized through the application of technology. The second will be dominated by innovation. Increasingly capable systems will transform the work of professionals, giving birth to new ways of sharing practical expertise. In the long run this second future will prevail, and our professions will be dismantled incrementally.*"

What happens to the social structures and social cohesion, if the Industry 4.0 fulfills its promises? Most European countries have adopted the welfare state model, which builds on high employment, high taxation, and active public participation in social, political, and economic life. In order to sustain the welfare state, industrial societies must maintain economic growth and high employment rates in both public and private sectors. In addition, modern professions are the foundation of modern societies. They foster social mobility and hold the middle-class in place. If modern professions collapse or disappear, the social structures will also collapse. There are already signs of political and social unrest which reflect the fears and frustrations of the middle-class [18].

The link between manufacturing systems and social systems is intimate and historically constructed. Changes in manufacturing systems have always generated social problems, which influence the political discourse. Daniel Buhr points to Elvis Hozdic and observes that the development of Industry 4.0 cannot be isolated from the social and cultural development of industrial societies. Internet, wireless networks, and the uninterrupted flow of information have already diluted borders and created a new social order, where the needs of individual citizens and customers are met by smart factories and flexible production methods. Individualization and the changing prospects of work put new challenges on the welfare state. According to the definition, a welfare state is supposed to counteract inequalities by redistribution and protecting against a set of risks. Industry 4.0 will produce new risks, which will penetrate in the very core of the welfare state. The welfare state is also based on social stratification, which more or less makes gainful employment a privilege. Again, the Industry 4.0 and digitalization puts this principal in jeopardy [19].

Hence, there are two ways to read the current visions of the future of manufacturing. Technological enthusiasm gets the most out of the technological promises that will bring a completely new concept to the industrial production. The Fourth Industrial revolution will bring a set of globally networked economic actors, who will reorganize and restructure the way work is conducted in post-industrial factories. The concept is simultaneously local, national, and transnational. Digital technologies and expanded Internet will allow companies to utilize global value chains and produce digitally manufactured goods to the global markets. Industrial structures are locally based, but they are designed to operate without local connections or regional expertise or labor market regulations. This model will create a globally standardized, networked production and service structures [12].

The other way to approach the Industry 4.0 is to place the concept in the historical context. This takes a relativist stand on the technological enthusiasm and puts light on the complex environment, where the new concept should operate. All industrial revolutions have promised dramatic changes and radical improvements in productivity. None of these visions have materialized, at least in the short period of time. As economic historians, based on the empirical evidence, have demonstrated, technological advances improve economic growth, but the improvements in productivity only come after long delay.

According to Paul David [20], the main cause to the productivity paradox be found in the manufacturing system itself. Factory owners and managers optimize production and they are reluctant to accept radical innovations, which will disrupt organization and working conditions. The other reason is found in the knowledge capacity, which is built into manufacturing systems over a long period of time. Skilled workers, managers and corporate leaders are unwilling to have to learn and adopt new knowledge and new methods—there is a lock-in to what is already known [21]. Hence, the diffusion of new knowledge and innovations slows down.

There is a rich literature explaining why innovations and new manufacturing systems don't break through the old systems. Without going deeper into the discussion, it is worth reminding that manufacturing systems are organic systems, which are managed by people. Although automated systems, robots, and digitalization will take over much of the routine tasks, the foundation of the systems will remain in the hands of skilled managers. Historical examples demonstrate that the change from old to new is always difficult and it involves a complex set of social and cultural factors. It is certain, that Industry 4.0 and the Fourth Industrial Revolution is coming. However, if the development follows historical examples, it is unlikely that Industry 4.0 will fulfill its strategic goals. Technological innovations will affect society, but society will affect technology. What comes out of this interplay is still an open question.

In vein with the above, this book is a collection of articles in the original spirit of the term "Industry 4.0", *focused on manufacturing* and presenting a holistic view of modern manufacturing. "Manufacturing 4.0" is presented, discussed, and analyzed in light of the technical, the economic, and the societal through visiting the past, the status quo, and imagining the future.

References

1. P. Temin, "Two views of British Industrial Revolution," *The Journal Economic History*, vol. 57, pp. 63–82, 1997.
2. J. Rifkin, *The Third Industrial Revolution. How Lateral Power is Transforming Energy, The Economy and the World*. London: Palgrave Macmillan, 2011.

3. S. K. Rao and R. Prasad, "Impact of 5G Technologies on Industry 4.0," *Wireless Personal Communication*, vol. 100, pp. 145–159, 2018.
4. R. Boyd and R. Holton, "Innovation, employment and power: Does robotic and artificial intelligent really mean social transformation?," *Journal of Sociology*, vol. 54, pp. 331–345, 2018.
5. D. F. Noble, *The Forces of Production. The Social History of Industrial Automation.* London: Routledge, 2011.
6. T. Hughes, "Evolution of Large Technological Systems," in *The Social Construction of Technological Systems*, W. Bijker, *et al.*, Eds., ed Cambridge, Mass., USA: MIT Press, 1987, pp. 51–82.
7. H. Kagermann, *et al.*, "Securing the future of German manufacturing industry. Recommendations for implementing the strategic initiative INDUSTRIE 4.0. Final report of the Industrie 4.0 working group," Plattform Industrie 4.0, Frankfurt am Main, Germany 2013.
8. J. Cowie and J. Heathcott, "The meanings of deindustrialization," in *Beyond the ruins: The meanings of deindustrialization*, J. Cowie and J. Heathcott, Eds., ed Ithaca, NY, USA: ILR Press, 2003, pp. 1–15.
9. J. H. Lorenzi and M. Berrebi, "The Impact of Deindustrialization," in *A violent world*, ed London: Palgrave Macmillan, 2016, pp. 74–101.
10. M. Banks and A. Ginrich, "Introduction," in *Neo-nationalism in Europe and Beyond: Perspectives from Social Anthropology*. M. Banks and A. Ginrich, Eds., ed New York, NY, USA: Berghahn Books, 2006, pp. 1–26.
11. World Economic Forum, "The Europe 2020 Competitiveness Report, 2014 Insight Report," Geneva, Swizerland2014.
12. S. Pfeiffer, "The Vision of "Industrie 4.0" in the Making—A Case of Future Told, Tamed, and Traded," *NanoEthics*, vol. 11, pp. 107–121, 2017.
13. E. Brynjolfsson, *et al.*, "Artificial Intelligence and the Modern Productivity Paradox: A Clash of Expectations and Statistics," 2017.
14. D. A. Hounshell. *From the American System to Mass Production, 1800–1932: The Development of Manufacturing Technology in the United States* Baltimore, MD, USA: Johns Hopkins University Press, 1985.
15. N. von Tunzelmann, *Technology and Industrial Progress: The Foundations of Economic Growth.* Cheltenham, UK: Edward Elgar Publishing, 1997.
16. N. Uddin and F. Hossain, "Evolution of modern management through Taylorism: An adjustment of Scientific Management comprising behavioral science," *Procedia Computer Science*, vol. 62, pp. 578–584, 2015.
17. R. Susskind and D. Susskind, *The Future of the Professions: How technology will transform the work of human experts.* Oxford, UK: Oxford University Press, 2015.
18. J. Schmid, *Wohlfahrtstaaten im Vergleich: Soziale Sicherung in Europa: Organisation, Finanzierung, Leistungen und Probleme.* Wiesbaden: Springer, 2010.

19. D. Buhr, "What about Welfare 4.0?," *CESifo Forum*, vol. 18, pp. 15–21, 2017.
20. P. David, "The Dynamo and the Computer: An Historical Perspective on the Modern Productivity Paradox," *The American Economic Review*, vol. 80, pp. 355–361, 1990.
21. M. Collan and F. Tétard, "Lazy User Model: Solution Selection and Discussion about Switching Costs," in *Nordic Contributions in IS Research*. vol. 86, H. Salmela and A. Sell, Eds., ed Cham: Springer, 2011, pp. 56–68.

PART I

Technical Perspectives

Additive Manufacturing—Past, Present, and the Future

Markus Korpela, Niko Riikonen, Heidi Piili, Antti Salminen, and Olli Nyrhilä

1 Introduction

Additive manufacturing (AM) is a relatively new manufacturing method that compiles different techniques to join materials together material on top of existing structure in order to make parts from 3D-model data—typically layer by layer. Additive manufacturing is a combination of

M. Korpela • N. Riikonen • H. Piili (✉)
Research Group of Laser Material Processing & Additive Manufacturing, Department of Mechanical Engineering, LUT School of Energy Systems, LUT University, Lappeenranta, Finland
e-mail: markus.korpela@lut.fi; niko.riikonen@lut.fi; heidi.piili@lut.fi

A. Salminen
Department of Mechanical Engineering, University of Turku, Turku, Finland
e-mail: antti.salminen@utu.fi

O. Nyrhilä
EOS Finland Oy, Turku, Finland
e-mail: olli.nyrhila@eos.info

M. Collan, K.-E. Michelsen (eds.), *Technical, Economic and Societal Effects of Manufacturing 4.0*,
https://doi.org/10.1007/978-3-030-46103-4_2

different technologies such as CAD (computer-aided design), CAM (computer-aided manufacturing), laser and electron energy beam technology, CNC (computer numerical control) machining, and laser scanning. Some of these technologies existed already in the 1950s, but only in the 1980s the maturity of the different technologies enabled the creation of additive manufacturing [1]. The term additive manufacturing substitutes historical terms, such as solid freeform fabrication, freeform fabrication, and rapid prototyping and it is also commonly called 3D-printing in non-technical contexts and in colloquial language [2, 3].

Additive manufacturing was originally developed around polymers, waxes, and paper laminates and used predominantly for prototyping purposes, as the term "rapid prototyping" indicates [3]. First commercial systems were available already in the 1990s [4]. Nowadays, most additively manufactured parts are functional and many of them are made of more advanced materials such as ceramics, composites, or metals [3, 5, 6]. AM techniques have developed rapidly, enabling AM parts to be used even in the most highly regulated application areas, such as in aviation. The increase of interest on AM has risen due to the speedy development of the technologies involved and due to digitalization. The initial fast advances in technology were followed by a hype, when the expectations about the capabilities of the technology were drummed-up through the social media and through various non-technical evangelists.

The first three industrial revolutions changed the world permanently and were the result of findings by inventors such as James Watt and Thomas Edison. Figure 2 shows how number of equipment connected to internet has been growing during various industrial revolutions. Additive manufacturing has often been connected to the 4th Industrial Revolution (IR4), which is claimed to have started after the year 2000. IR4 is merely about the digitalization and the networking of various technologies. The IR4 is a result of various technologies being mature for new tasks around the same time and the fact that high speed data-transfer and enormous (compared to previous times) computer capacity is available. The major technologies involved in IR4, like augmented reality, additive manufacturing, and artificial intelligence are studied by various companies ranging from retail to manufacturing and from transportation to banking [7]. European Patent Office made a study about patents and IR4 and found that the number of patents filed related to the IR4 increased with 54 percent in the past three years [8]. The number of patents in additive manufacturing has grown from few applications in 1991 to a "five hundred per year"-level in 2015, the growth in the number of patents is approximately 150 per year. Interestingly, only a fraction (3.5%) of the filed patents were

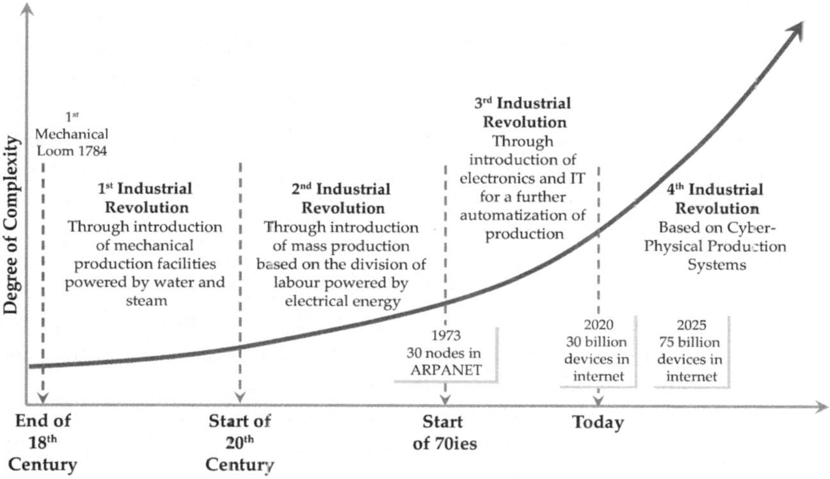

Fig. 1 The past and current industrial revolutions from the 1st to the 4th, and the number of devices connected to the Internet [10–12]

directly connected to IR4 main constituents—digitization and networking [1]. The IR4 and so-called smart factories are considered to be crucial game changers for the survival of US and European industries because they have the ability to reset the labor productivity back on the growth-path again [9]. The past and current industrial revolutions from the 1st to the 4th, and the number of devices connected to the Internet is presented in Fig. 1.

Main advantages of AM are connected to the ability to manufacture complex geometries [13], lighter structures, and the ability to allow customization. In subtractive manufacturing processes, an increase to the volume of material removed from the billet or an increase in the geometric complexity of the design cause the manufacturing time to be longer and the manufacturing costs to be higher. This often leads to components having excess material that cannot be cost-effectively removed [14]. Often the majority of removed material in conventional manufacturing ends up as waste [15], of which most can be recycled, but the value of this waste is typically just a fraction of the value of the original material. In additive manufacturing processes, complexity itself does not add costs in the same way and the material is added primarily only to where it is needed—thus parts are lighter by default. As an automated manufacturing process, AM shares similarities with CNC machining. It cannot currently add material faster than high-speed CNC machining can remove it, but it can manufacture more complex parts in a single-step process within a certain framework [3]. As a limitation, AM suffers from surface integrity on specific surfaces and of a degraded dimensional control [16], which often leads to the need of post-processing, especially in metal additive manufacturing. Attempts to solve this problem have been made via using hybrid systems that combine AM and CNC machining, but utilization level of these systems is typically very low in the industry (Fig. 2).

The field of additive manufacturing is suffering from a lack of standardization, because most already existing standards cannot be utilized in AM [3, 4], furthermore the technologies are developing so rapidly that standardization cannot keep up. The already existing non-AM standards rest on known homogeneous microstructures of standard bulk materials. This does not apply to AM, in which the microstructure, created during the process can vary within a part and depends on the used process parameter

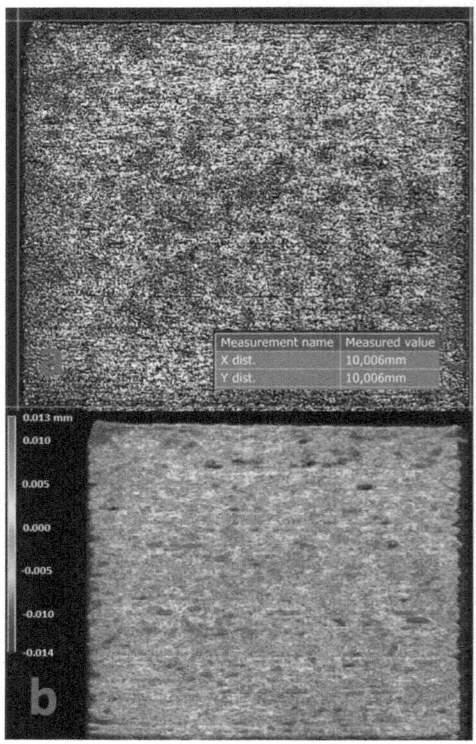

Fig. 2 Macro images of an additively manufactured (L-PBF), 10 × 10 × 10 mm³, tool-steel cube (a) showing geometrical inaccuracy, and (b) showing height differences on the surface and distortion in the left upper corner of the side of the cube

values. Tens, or possibly more than a hundred, different national and international AM standards have been approved, but additive manufacturing as a manufacturing method consists of thousands of different materials from polymers to metallic shape-memory alloys. As a reference, it took about 20 years to publish main European standards related to only aluminum in the context of conventional manufacturing [17].

2 Additive Manufacturing Materials and Processes

It was estimated in 2004 [18] that exist 40 000 to 80 000 engineering materials, and at least a thousand different processing methods for them. Just to point out, more than 20 000 different steel alloys are available [19]. The range of materials available for AM is only a fraction of the total range [20] despite the high speed of material development. The development of a new material is feasible, if an application or the potential for a new material has been recognized and evaluated such that it covers the material development cost. In AM, and especially in metal AM, absolute material costs are low due to low production volumes and therefore savings in material costs play a relatively minor role. This slows down materials development, but the narrow material-repertoire is caused also by the short history of AM. Most of the research so far has been concentrating on the manufacturing of existing and approved materials with the idea to make them also suitable for manufacturing with AM [21]. The basic material development in AM usually means that the process is defined such that reliable production with a material can be guaranteed. The current development of AM materials is typically based on the use of conventional alloys that (already) exist for traditional manufacturing [22]. About 2800 different commercial AM materials are available today [51]. It is notable that commercial materials include "multiples of the same materials" with a different names or brands by different manufacturers. For example, there are multiple choices for stainless steel AISI 316L as each manufacturer has their own name for it.

The materials development in AM can be seen in light of two different approaches—in the first approach an AM process for a traditional engineering material is developed such that the AM material properties correspond to traditional material properties as closely as possible—in the second approach the characteristics of an AM process are utilized to produce such materials that are only available for AM. For example, there is a possibility to create new characteristics for engineering materials, such as an optimized micro-structure for metal materials in a powder-bed fusion (PBF) process [10, 22–24]. The development of AM specific alloys shows the interesting potential of AM for the future. The aluminum-alloy "Scalmalloy" is an example of such a developed alloy for which the mechanical properties are enhanced to fit the typical thermal cycle of laser based powder bed fusion (L-PBF) [25, 26].

Systematic knowledge about the properties of metal AM parts is missing and the repertoire of available materials is still limited [23, 27, 28] mainly to different stainless- and tool-steels, aluminum- and titanium-alloys, nickel-based superalloys, and cobalt-chromium alloys, consisting of about 30 different materials in total [29]. A study of Herzog et al., [21] covered 159 references of metal additive manufacturing with PBF and directed energy deposition (DED) processes. Their comprehensive analysis resulted that static and fatigue strength together with other mechanical properties of AM materials are alike to their conventional counterparts and that AM metals and alloys can be evaluated by known concepts of fracture-mechanics [21]. Copper and some copper alloys have already become available, but are unfortunately not on the same maturity-level as the above-mentioned materials, due to their material property related issues such as low absorption of current laser beam wavelengths and especially high thermal conductivity, [30]. As non-engineering materials, some precious metals, such as gold and silver, are available as well. Similar material-databases that are available for conventional materials do not exist for additively manufactured materials, and the properties of the printed materials are neither discussed deeply in recently published books in the field of metal additive manufacturing [3, 27, 31, 32]. AM materials from the material research point of view are discussed deeper elsewhere in this book.

Tens or even hundreds of different AM-techniques exist, but EN, ISO, and ASTM approved standards categorize them into seven different process-categories. The process categories are listed in Table 1.

In metal AM, directed energy deposition (DED), powder bed fusion (PBF), and sheet lamination are single-step processes in which basic material properties, such as density of more than 90%, are achieved in a single operation step. Material extrusion and binder jetting of metal parts are multi-step processes in which the parts require consolidation by a secondary process such as sintering in an oven, in order to result density greater

Table 1 Process categories of additive manufacturing, with definitions according to EN ISO/ASTM 52900 [2]

Process category	*Definition:*
	"An additive manufacturing process in which ..."
Material extrusion	*"... material is selectively dispensed through a nozzle or orifice"*
Powder-bed fusion	*"... thermal energy selectively fuses regions of a powder bed"*
Binder jetting	*"... a liquid bonding agent is selectively deposited to join powder materials"*
Vat photopolymerization	*". . liquid photopolymer in a vat is selectively cured by light activated polymerization"*
Material jetting	*"... droplets of build material are selectively deposited"*
Directed energy deposition	*"... focused thermal energy is used to fuse materials by melting as they are being deposited"*
Sheet lamination	*"... sheets of material are bonded to form a part"*

than 90%. Material jetting of metal parts is a rare production method, but at least one system manufacturer has established a machine for that purpose. According to the manufacturer, their process is a single-step process[33]. Different kinds of applications of different additive manufacturing processes can be seen in Fig. 3 [2, 27].

Out of these seven process categories, powder-bed fusion has shown the highest potential in additive manufacturing of metal parts, and in more detail, the PBF process that utilizes laser beam as a heat source [32]. It is also the most studied AM technique [16].

3 Metal Additive Manufacturing

Metal additive manufacturing is a over 30 years old manufacturing method [34] which has now grown to a point, where it is a potential method of manufacturing for real-world applications. Parts manufactured with the most common and widely applied metal additive manufacturing processes are only semi-finished, but in some cases, they can be used directly as end-products [32, 35, 36]. Metal AM enables building of geometries that conventional subtractive manufacturing is not capable of [53].

Metal AM is still a niche market in manufacturing, but it is growing at a fast pace. Current systems of metal additive manufacturing are not something to completely revolutionize way of manufacturing, or to completely replace traditional manufacturing methods. AM is an addition to the repertoire of manufacturing methods and more likely will replace other

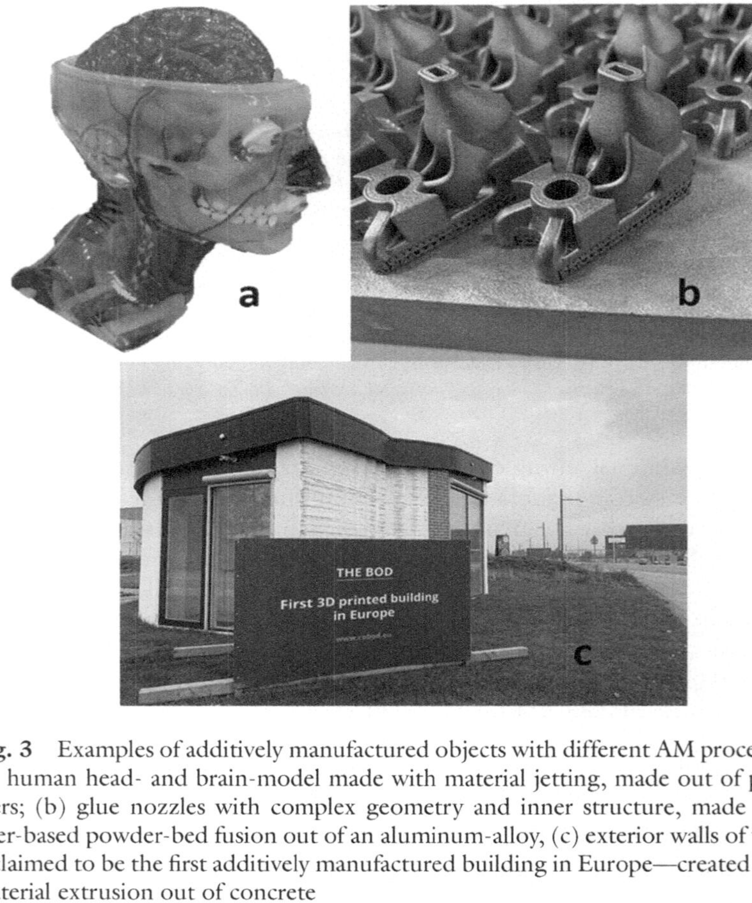

Fig. 3 Examples of additively manufactured objects with different AM processes; (a) human head- and brain-model made with material jetting, made out of polymers; (b) glue nozzles with complex geometry and inner structure, made with laser-based powder-bed fusion out of an aluminum-alloy, (c) exterior walls of what is claimed to be the first additively manufactured building in Europe—created with material extrusion out of concrete

manufacturing methods or manufacturing steps for *certain applications*. Despite the hype around additive manufacturing and its possibilities, metal AM has many restrictions that make it the most convenient choice of manufacturing for only a limited number of applications. This is due to the rather high costs (often >1000 €/kg for end-usable AM steel part) incurred by the required post-processing and the slow manufacturing speed of the rather expensive systems that need to be used. The applications that are economically viable to create by metal AM are typically (but not limited to) the size a human fist and geometrically (very) complex and if conventionally manufactured would require multi-step manufacturing processes such as machining, assembling, or joining. On the other hand, utilizing AM is beneficial for parts that have a geometry that typically requires casting, but for which the production volumes are low and would thus lead to rather high mold costs divided per unit—AM can be used to manufacture parts so complex that several casted parts would have to be joined to otherwise create them.

The layer-by-layer manufacturing method is not a new method for creating metal parts. It has been possible to cut holes to metal sheets and join them together via welding or by using fasteners for decades. This enables the same advantages as AM, such as possibilities to create complex geometries inside of a part. However, these kinds of sheet laminated applications have been very rarely utilized despite the opportunities they offer.

Despite the limitations and high expenses, utilization level of additive manufacturing is presumable lower than it could be. Certain parts, originally designed to be manufactured with subtractive manufacturing methods, could be more cost-effectively manufactured with AM, especially after redesigning these parts for AM—often such parts are still manufactured with conventional methods, lack of AM-knowledge might be one reason for the non adoption. Identifying parts suitable for AM requires lot of knowledge about the advantages and the disadvantages of metal additive manufacturing and experience about how to apply AM into industrial cases. This knowledge is not part of current engineering curricula so most engineers in the industry are not aware of it. Therefore, some companies have no knowledge about additive manufacturing, whilst some use it daily.

4 Powder Bed Fusion

This article deals mainly with laser based powder bed fusion, because it is the most common, widely applied, and possibly the most evolved metal additive manufacturing technology available [27, 32]. It is also the most used metal AM technology for production of engineering components [27].

Powder bed fusion is based on melting metal powder to form parts layer-by-layer. The melting is based on melting of powder with electromagnetic radiation that is photons from a laser, or with electrons from an electron beam (EB). The laser beam is absorbed via Fresnell absorption, or plasma absorption, by the material both being relevant in the PBF process, whereas in electron beam melting EBM the kinetic energy of electrons is transferred into heat during the interaction with the powder material. Both can be used for manufacturing high precision parts. However, electron beam-based systems are rare and therefore not thoroughly introduced here.

The laser beam is focused and guided on the surface of a metal-powder-bed, which is inside of a building (manufacturing) chamber, filled with inert gas (L-PBF), N or Ar, or a vacuum (EB-PBF/some L-PBF systems). Once the predefined areas of the powder-bed are melted, a machine-integrated recoater, or roller, will automatically spread another thin, typically 20–90 μm, powder layer on top of the previous one and the building platform goes down (is lowered) by as layer thickness and the process is repeated. The energy density of the laser beam must be high enough for sufficient melting and making the melt pool deep enough to reach the previously melted layer (solid). Some part of the beam reflects away from the powder bed, whilst a large part of it is absorbed by the material and melts it. The absorption of the powder is significantly higher on the powder bed than on a flat surface of solid metal, due to multiple reflections of the beam and large specific surface area of the powder material [23, 32]. When the beam moves on the bed, molten material is solidifying "behind" the beam and cooling down. Since the thermal cycle is very fast, also the solidification and cooling rates are high [37, 38]. High cooling rate applies to L-PBF and causes a significantly different microstructure from the conventionally manufactured counterparts to form [23]. Imperfections, such as undesired microstructures, high residual stresses, and porosity can occur in powder-bed fusion [28]. During the process, at any particular time of melting, some parts of the work-piece are contracting during cooling, while others are expanding when heated. Cyclic heat delivery is inherent in L-PBF and it induces residual stresses to the parts, because of subsequent thermal expansions and contractions. These residual stresses can rise so high that eventually they can lead to bending and distortions within the part [32, 39].

Many system producers use their own commercial names for PBF such as selective laser sintering (SLS), direct metal laser sintering (DMLS), selective laser melting (SLM), laser cusing, or electron beam melting (EBM). Despite the word "sintering" in the names of some systems or processes, the current *metal* PBF systems completely melt the particles instead of sintering them [32], whereas in the PBF of polymers, the particles are either fused or sintered [40]. The number of different PBF system-producers is more than thirty. The system-producers and their machine *base* prices are published in an annual report by Wohlers Associates [24]. The average prices for these systems are presented in Table 2. However, *total* investment costs for a state-of-the-art mid-size L-PBF system and the required auxiliary equipment is closer to one million, than half a million euro.

L-PBF has developed a lot during the last twenty years. In the early 2000's, the machines started to be equipped with new, at the time just matured, fiber lasers. The efficiency of a fiber laser is much higher than that of the previously used CO_2 lasers, which increased resolution and accuracy of the process remarkably. This was a major improvement, based on the improvement of beam quality, availability of cheaper and higher quality optics, simpler arrangement of the optical path, and the improved absorptivity of the laser beam to the metallic material, all enabled by utilization of fiber laser. In the late 2000's, the power of the lasers in L-PBF systems was increased to some hundreds of watts, and in the early 2010's, multi-laser systems were introduced. The number of lasers used and their power have both increased since then. Most of the models of different system-producers operate with one to four lasers in within the 100–1000 watt power-range. The cost of an additional laser in a system is typically less than the benefit achieved by improvement of productivity. According to the web-pages of four large L-PBF system producers, the production-speeds of their flagship models are between 100–171 cm^3 per hour. The machines are equipped with two or four 400, 500, or 700 W lasers [41–44]. Parameters guaranteeing the highest production-speeds are not disclosed and therefore it might be that these values do not correlate with normal manufacturing speeds or with best achievable accuracies. For example, increasing layer thickness has a major positive effect on building time and negative effect on accuracy. Therefore, an unambiguous value for

Table 2 Calculated average maximum build volume and average prices for L-PBF systems, based on data available in annual report by Wohlers Associates [24]

L-PBF system	*Build volume (liters)*	*Average build volume (liters)*	*Average price (K€)*
Average system	20	–	~500
Small system	<10	1.5	~200
Medium system	10–30	20	~400
Large system	>30	70	~1000

the volume rate, or the cost per kilogram cannot be stated due to "problem complexity"—the same applies for machining. However, a rough idea of the costs can be introduced. Volume-rate of one of the most common state-of-the-art mid-size AM system is approximately 15.2 cm^3 per hour for one of the most common steel materials [45]. Diegel et al. [46] estimates that a 650,000 USD investment on an AM system leads to an hourly running cost of 48.69 USD. According to these numbers, a hourly machine running cost of constructing a theoretical one liter of solid steel would alone be ~3200 USD via L-PBF AM. This result to about 400 USD/kg, but it needs to be noted that this cost excludes, for example, the margin of the manufacturer, the labor, and the materials-related costs.

As mentioned, PBF is possibly the most evolved AM technology, but the production speed is still slow and the production is expensive compared to conventional manufacturing—furthermore the parts are usually semi-finished products that require post-processing [32, 35, 36]. However, the systems are developed constantly and new system-integrators are appearing on the market. The traditional system vendors are also coming to the additive manufacturing markets with new concepts. The competition is getting tougher since these traditional vendors have a consistent existing customer-base among the manufacturing companies.

5 Other Metal Additive Manufacturing Technologies

Directed energy deposition (DED) is another important method for metal additive manufacturing. The principle of the method is totally different from powder-bed fusion. The method is based on dynamically feeding material into the molten pool created with a heat source. The material can be in form of powder or wire, whereas the heat source to melt material is a laser beam, an electron beam, or an electric arc. The processes have their own descriptive names under the category of DED by ASTM. The laser-based process is typically called Laser Metal Deposition (LMD), electron beam-based Electron Beam Melting (EBM), and arc source-based Wire Arc Additive Manufacturing (WAAM).

The major differences between DED processes and the PBF are accuracy, building speed and build volume. Accuracy being better in case of PBF, whereas the building speed and the build volume are major advantages of the DED processes. DED processes are also typically capable of handling much higher heating power than PBF. DED processes can utilize high power with high building speed; typical laser power used is 1–4 kW. The major difference of DED compared with PBF is the size of the part, which in case of PBF is restricted by the size of building chamber (typical sizes are reported above). Building volume can be several meters in each direction in case of DED, depending on the dimensions of the working area of the robotic system used. The accuracy on the other hand is typically around 0.1 mm for robotic systems and especially in the case of WAAM the accuracy is considerably lower.

The surface quality of DED parts is of a lower standard than that of the PBF process. This is typically, because of larger molten pool volumes and the dynamically fed material. In case of WAAM there is typically a lot of spattering leaving spatters on top of previously build parts. In case of powder LMD some of the powder is hitting the solidifying surface of melt pool and sticking there and melting only partially. In order to reach typically accepted part surface quality levels, all visible surfaces must be machined prior to further use. The utilization of DED is increasing at a considerable speed among manufacturers of larger components, where they have some specific areas of application, for example, in building shapes on top of a sheet structure to reach some of the advantages in design with reasonable pricing. Similar case is especially true with WAAM technology, which has appeared only during the last 2–3 years, even though the technology has been available for tens of years. WAAM techology is predicted to make only some percentages of the metallic printing market, but many conventional companies are interested in it, because they are already familiar with the technology from the perspective of welding.

Other metal AM methods exist, but they are relatively new processes, without many existing system-producers and without a significant number of scientific results published about them. Material extrusion of metals has expanded a lot during the last couple of years, but cannot produce as high-quality parts as powder-bed fusion is capable of producing. The same applies to binder jetting of metal parts. Parts made with these processes have a remarkable lack in density, which leads to worse mechanical properties. On the other hand, total investment costs of both methods are lower than in the case of powder bed fusion.

6 Technology Readiness Level

The entrance of a new technology to an existing technical environment is typically difficult. In practice, the new technology should be better and cheaper than existing technologies that have proven their position in the market. This is typically true within manufacturing, where there are often tight requirements in respect of mechanical properties and performance for part manufacture. The entrance of a new manufacturing technology is often very difficult because of this. Thus, there have not been real newcomer technologies, unless of the laser-based AM technologies are counted. The validation of a new technology is typically based on the existing technologies. A typical way to evaluate the state-of-the-art and capability of technology is assessment of so-called Technology Readiness Level (TRL). This evaluation scale is originally developed by NASA and is currently spreading to new applications. TRL scale is nine steps from introduction of technology to final accepted level where full utilization of technology can be carried out [1, 48]:

1. Basic manufacturing implications identified.
2. Manufacturing concept identified.
3. Manufacturing proof of concept developed.
4. Technology validated in laboratory environment.
5. Basic capabilities shown (near production environment).
6. System produced (near production environment).
7. Production in production environment demonstrated.
8. Pilot line capability demonstrated.
9. Low rate production.
10. Full rate production.

Because of the constant development-work carried out in technology and application level, various AM technologies and applications are in different TRL-levels. Some of the applications are already in TRL level 10, this includes applications within the aviation industry. An example of how a TRL-process goes forward is heat exchangers developed for NASA using Ultrasonic Additive Manufacturing 3D metal printing and elevated Ultrasonic Additive Manufacturing (UAM), a sheet lamination process. The company "Fabrisonic" reached TRL level 3 (proof of concept) and then TRL level 6 (prototype demonstrated in relevant environment) in

2018. The company has successfully completed all tests required by NASA JPL for flight qualification [49].

7 On the New Possibilities AM Offers

Additive manufacturing enables multiple benefits in manufacturing parts compared to traditional manufacturing methods. Due to the layer-per-layer manufacturing process, AM allows the placement of material only to the areas required, which saves material and results in lighter parts [3, 31]. By using advanced design and modeling software this freedom can be highly utilized due to the possibility to design complex AM parts that are as strong as, or even stronger than traditionally manufactured parts [24, 27].

In engineering, parts are often joined to form a larger assembly. AM reduces the need to join parts, as the joints can be integrated into the part. This means that complex piping can be built in a single piece without the need of making separate pipes and joining them together later by, for example, welding or mechanically by bolts. However, AM does not necessarily remove the need to join parts, because of the limited building volumes available.

AM opens new possibilities to tailor products for users. In general, AM is a single-step manufacturing process in which the part is built layer-by-layer typically on an empty platform with one machine, and thus the effort put in the manufacturing process itself is minimal compared to multi-step manufacturing methods such as casting and multi-round machining [24]. Due to the single-step process, small batches or single parts can be made more effectively than with conventional manufacturing methods. Also, the higher degree of freedom in AM compared to conventional manufacturing methods enables more unique shapes and products to be made [24, 31, 47].

AM is well-suited to allow the optimization of flow and heat characteristics of parts. Enhanced gas or fluid flow properties can be achieved, for example, with the possibility to manufacture smooth cavities with no sharp turns or corners. Enhanced heat-flow properties can be achieved by making large surface areas to, for example, heat exchangers [24, 31].

AM provides an opportunity to reduce the number of agents in a logistic chain, because a part can be printed with one machine without the need of molding or tooling. This is beneficial especially in the cases, where the part is printed in a single-step AM process and does not require further processing. AM also makes it possible to manufacture spare-parts for which the original tooling, molds, or production machines no longer exist [24, 27, 47].

The future of AM materials is predicted to experience a noteworthy change. In case of metals, the process is totally different from the conventional steel making and features like local metallurgy are going to provide completely new alloys [22]. The current selection of available materials reflects the needs of current optimized metal manufacturing processes and there is a lot of potential for new commercial materials specific to AM. One possibility, which is currently studied in various locations, is that the small molten pool size typical for the PBF process enables much wider use of alloying elements than traditional manufacturing [10]. This can lead to an era of totally new materials with new sets of properties. AM differs from conventional manufacturing in the way that it does not utilize bulk material of which he geometry is modified, but also enables the modification of the microstructure. This means that microstructure is basically created during the printing process and AM equipment and systems do it differently [24]. For example, AM process-parameters and location of a part in building platform have a huge effect on cooling rates for any geometry. In conventional manufacturing, microstructures of parts are not changed during machining, but they can be controlled and modified by different standardized heat treatments performed after the processing. In AM, different heat treatments are part of the manufacturing process itself, but also separate heat treatments are applied to parts afterwards. One possibility of future of AM is the ability to control the heat treatments during the process to produce desired microstructure on demand [24].

8 Challenges for AM

Additive manufacturing has been said to completely revolutionize manufacturing or even being the fourth industrial revolution. The said manufacturing revolution of AM would unfortunately require that any components in any geometry could be manufactured in a single-step manufacturing process requiring no, or lower operator skills, and no assembly and resulting in a more cost-effective solution than the current manufacturing processes. AM can already produce some end-use metal components in a single-step process, but the area of these applications is very narrow and is limited mainly to prototypes. The restrictions come from the already mentioned surface quality and the degraded dimensional control that do not have a large role in prototyping. The dimensional control can be seen as manageable, once AM related software are developed enough, but there is no technology on the horizon for solving the surface quality issues in a single-step manufacturing process without adding other technologies to AM systems. In many goods, a lower surface quality would be sufficient enough, but "unfortunately" certain surface quality-levels have already been adopted as standard. On the other hand, most goods require better surface quality than what can be achieved via single-step additive manufacturing process. Worse surface quality means a larger surface area and that in turn means, for example, a larger area for bacteria and other impurities to fasten to and makes their removal harder. In mechanical engineering, certain tolerances are unavoidable and unfortunately outside the achievable scope of the dimensional accuracies of current additive manufacturing systems—later on higher accuracies become possibly achievable via more developed software that can predict output geometries better and scale models accordingly. At this stage, additive manufacturing of parts that can meet high tolerances require multiple iteration steps that result in inefficient cost structures. A lot of effort has been put on AM production of spare-parts, but unfortunately mentioned restrictions of the process makes many of these projects only conceptual studies without economic viability. However, a competitive company must stay on top of technology development and in order to stay competitive one must be active already when a technology reaches the breakthrough-point—not after.

To date it is hard to state that AM would be more than an addition to the repertoire of available manufacturing methods. One can wonder whether AM could revolutionize manufacturing in the future and there is some truth to that kind of thinking, but at this stage there is no technology on the horizon to solve the already mentioned restrictions related to the AM process itself. In the 1990's, it was relatively easy to predict the future importance of the Internet as only the infrastructure was missing. The technology was already there but was expensive. For example, in metal additive manufacturing, the required technology is not here yet, nor is it on the horizon. In theory, a machine that could combine many different manufacturing methods leading to a machine that can build almost anything in a single-step process, would be the solution, but AM will probably not have major role in such a machine.

The most advanced metal AM processes are based on micro-scale laser welding and have the same limitations as "conventional" laser welding.

Laser welding is highly studied and utilized in the most advanced manufacturing industries, but still cannot have welding speeds higher than what is specific for certain resolution due to dynamics of melt pools. The same applies to metal AM; the process speed of the laser cannot be increased without limits and are already operating at limit, as we know them. In the same way, the power of lasers cannot be increased limitlessly, because the higher the power density the lower the resolution. Number of lasers used can, and has been increased and multi-laser system have been on the market for a couple of years already, but single laser systems are still market leaders.

Another issue is that many different quality of products can be additively manufactured. The most common machines are the so called low-cost consumer 3D-printers, of which producible parts-quality does not represent the quality that an industrial high-end AM system is able to produce. Some people see only AM parts made with these low-cost machines and seem to have the notion that they are reflective of the capabilities of the additive manufacturing technology as a whole. These kinds of wrong impressions tend to slow down companies' adoption- and utilization levels of additive manufacturing.

A very common statement in the field of AM is that AM has no limitations in geometry or that *design freedom* exists in AM. It is possible to produce geometries that for example are not possible with CNC machining, inner structures are a good example of these. But the statement is only partially true, because on the other hand, AM, and especially metal AM lacks the possibility to create even some of the simplest geometries, like precise cubes and balls, or the so-called unsupported geometries. Cubes include sharp corners that have stress concentration and when a cube is large enough the too-large surface to be melted may lead to cracking of the metal. Ball shaped structures includes unsupported areas and if large enough are impossible to build without separate support structures and without heat sinks used to dissipate the heat away from the largest surface areas (to avoid excessive distortion caused by narrow contact area to the building platform). Hollow structures always mean that unsupported overhanging structures exist, which leads to the unavoidable need of support structures. Removal of support structures, for example, from inside of a hollow cube is impossible without breaking the structure. In addition, unmelted powder remains inside hollow structures and needs a separate hole for removal. Creating of these holes, separate heat sinks, or support structures is not a problem from the manufacturability perspective, but they often increase the need for post-processing. Additional support, or heat sink structures, mean that extra metal is welded to the part and needs to be removed often leaving undesired rough surfaces to their original locations. All these additional, but in many cases mandatory, steps take metal additive manufacturing further away from being a single-step manufacturing process, which is said to be its main advantage. This means that the statements "unlimited design freedom", or "possibility to create any geometries" have a truth-value of less than one. One limiting factor in the widely discussed freedom of design in AM is also the available building volume [47]. For example, metal powder-bed fusion machines have an average building chamber volume of approximately 20 liters. Larger building volumes are available, but they are rare because of remarkably higher investment costs involved, which lead to even higher machine hourly running costs [24]. Increased productivity covers the excess costs only, if the building volume is fully utilized in each manufacturing run. As mentioned, the PBF technology is a remarkably slower manufacturing method compared to conventional methods, as it is not capable of adding material at the same speed as machining is capable of removing it from a solid work piece [3]. Due to the slow building speed, the machine running costs usually climb high especially in tall builds. A build taller than 200 mm can take more than a hundred hours to construct.

9 Future Trends and Development

Regarding AM technology, the major trends are increasing the building speed, freedom in design, and the level of automation [47]. For example, the current use of a typical mid-size, mid-cost metal L-PBF machine requires a lot of manual work for filling and unpacking powder and for loading and unloading the building platform. Also moving the building platform from the printer, for example, to an oven for thermal treatment, and from there to a band saw for part removal is not a comfort for the machine operator. Support structure removal is currently a manual process, especially in connection with complicated parts that have support structures in locations unreachable with CNC tools [24]. There are existing projects that aim to replace the manual working phases with automation. The filling, unpacking, and recycling of the powder used is automatically done in some machines. Also, systems that have integrated thermal treatment within the machine are available. In the future, the automation level can be expected to spread into all manufacturing phases in such a way that the role of the machine operator will become that of supervisor [47].

Building speed is a major limitation of the L-PBF process and system-manufacturers are constantly developing methods to increase it. For example, multi-laser systems, multi-recoater systems, and multi-direction recoater systems have been used to increase the building speed. However, the building speed is not yet satisfying for all needs, because AM is "always" compared to other manufacturing methods that are remarkably faster in most cases [47]. In some AM technologies such as vat photopolymerization, the building speeds have been increased to a level in which parts can be built in minutes.

Also new technologies arise within the AM realm. For example, the so called cold spray technology aims to tackle the limitations of building speed in metal AM by spraying metal powder particles with high speed to the desired locations in order to form a 3D-part. When the particles hit the building platform or the previous layer of the part, their kinetic energy transforms into heat and the particles melt on top of the previous layer [50]. This technology is claimed be 100–1000 times faster than the traditional metal AM.

Improvements in post-processing of metal parts are also being developed. The required support structures in metal L-PBF are a major restriction for the freedom of design [24], and thus, systems that focus purely on post-processing, are developed. For example, there are systems that are removing support structures automatically without mechanical work. The process works by "electrochemical pulse methods, hydrodynamic flow, and particle assisted chemical removal" according to one manufacturer. The problems with the support structures are also being solved from another point of view—to reduce the volume of needed support structures. New technology improvements allow the construction of parts in lower angles without support structures, but this does not eliminate the need of support structures completely; currently the parts must be at least anchored to the building platform [24, 31].

The quality of parts, especially metal parts, is a hot topic in AM. Process monitoring can be found from many L-PBF machine manufacturers' machines nowadays, but they are being developed for better performance to meet the quality assurance desires. The future trend is to develop systems, where process monitoring can detect a fault in the build and react to fix it during the next few layers. Since the nature of the PBF process is based on melting also some of the previous layers, corrective actions can be taken during at least two layers that follow the layer with the fault. This opens a window of opportunity to react to flaws in time, if sensors and algorithms are developed to be good enough. These so-called adaptive systems, which are not yet available, could really remove the flaws during the build [51, 52, 53].

References

1. NASA—National Aeronautics and Space Administration, (2013)—NPR 7123.1B—NASA Procedural Requirements—Systems Engineering Processes and Requirements, USA, 2013.
2. ISO/ASTM 52900:2015 Additive manufacturing — General principles — Terminology.
3. Gibson I, Rosen D, Stucker B, (2015) Additive Manufacturing Technologies: 3D Printing, Rapid Prototyping, and Direct Digital Manufacturing. Second Edition. New York: Springer. 498 p.
4. Monzón M, Paz R, Ortega Z, Diaz N, (2019) Knowledge Transfer and Standards Needs in Additive Manufacturing. In: Additive Manufacturing—Developments in Training and Education. First edition. Cham: Springer International Publishing. 228 p.
5. Wohlers T, Campbell I, Diegel O, Kowen J, Fidan I, Bourell D L, (2018) Wohlers report 2018. 3D printing and additive manufacturing state of the industry annual worldwide progress report. 342 p.
6. Tofail S, Koumoulos E, Bandyopadhyay A, Bose A, O'Donoghue L, Charitidis C, (2018) Additive manufacturing: scientific and technological challenges, market uptake and opportunites. Materials Today 21(1): 23–37.
7. Schulze E, (2019) Everything you need to know about the Fourth Industrial Revolution. CNBC. Internet document. Cited: 27.1.2020. https://www.cnbc.com/2019/01/16/fourth-industrial-revolution-explained-davos-2019.html.
8. European Patent Office, (2017) Patents and the Fourth Industrial Revolution: The inventions behind digital transformation December 2017. Internet document. Cited 27.1.2020. Available: http://documents.epo.org/projects/babylon/eponet.nsf/0/17FDB5538E87B4B9C12581EF0045762F/$File/fourth_industrial_revolution_2017__en.pdf.
9. Wellener P, Shepley S, Dollar B, Laaper S, Manolian H A, Beckoff D, (2019) 2019 Deloitte and MAPI Smart Factory Study. A report from Deloitte's research center for energy & industrials group. Deloitte. 21 p. Cited 30.1.2020. https://www2.deloitte.com/content/dam/insights/us/articles/6276_2019-Deloitte-and-MAPI-Smart-Factory-Study/DI_2019-Deloitte-and-MAPI-Smart-Factory-Study.pdf.

10. Poprawe R, (2018) Digitale Photohiche Production—Die Zukunft der Laser-Materialarbeitung. Proceedings of 18th Laser Technology Congress AKL, Aachen, Germany, May 2th–4th 2018, Fraunhofer Institute for Laser Technology.
11. World Economic Forum, (2019) The Fourth Industrial Revolution: what it means, how to respond. Research Agenda. Cited: 30.1.2020. https://www.weforum.org/agenda/2016/01/the-fourth-industrial-revolution-what-it-means-and-how-to-respond/.
12. Statista, (2019) Internet of Things (IoT) connected devices installed base worldwide from 2015 to 2025. Cited 30.1.2020. https://www.statista.com/statistics/471264/iot-number-of-connected-devices-worldwide.
13. Niaki M K, Nonino F, (2018) The Management of Additive Manufacturing: Enhancing Business Value. Cham: Springer International Publishing. 220 p.
14. Muthu S, Savalani M, (2016) Handbook of Sustainability in Additive Manufacturing, Vol. 2. Singapore: Springer. 114 p.
15. Huang R, Riddle M, Graziano D, Warren J, Das S, Nimbalkar S, Cresko J, Masanet E, (2016) Energy and emissions saving potential of additive manufacturing: the case of lightweight aircraft components. Journal of Cleaner Production 135(1): 1559–1570.
16. Townsend A, Senin N, Blunt L, Leach R K, Taylor J, (2016) Surface texture metrology for metal additive manufacturing: a review. Precision Engineering 46: 34–47.
17. Raaka-ainekäsikirja 5: Alumiinit, (2002) Tampere: Metalliteollisuuden Keskusliitto, MET. 235 p. Raaka-ainekäsikirja -sarja. MET-julkaisuja nro. 13/2002. [Handbook of Raw Materials 5: Aluminums] (in Finnish).
18. Ashby M F, Brechet Y J M, Cebon D, Salvo L, (2004) Selection strategies for materials and processes. Materials & Design 25(1): 51–67.
19. Miekk-oja H M, Lindroos V, Sulonen M, Veistinen M, (1986) Uudistettu Miekk-ojan metallioppi. Teknillisten tieteiden akatemia: Otava Helsinki. 841 p. [Renewed edition of science of metals] (in Finnish).
20. Smith P, Rennie A, (2010) Computer aided material selection for additive manufacturing materials. Virtual and Physical Prototyping 5(4): 209–213. https://doi.org/10.1080/17452759.2010.527556.
21. Herzog D, Seyda V, Wycisk E, Emmelmann C, (2016) Additive manufacturing of metals. Acta Materialia 117: 371–392.
22. Weisheit A, (2018) Materials engineering for laser additive manufacturing. Proceedings of 18th Laser Technology Congress AKL, Aachen, Germany, May 2th–4th 2018, Fraunhofer Institute for Laser Technology.
23. Sun, S., Brandt, M., Easton, M., (2017) Powder bed fusion processes: an overview. Laser additive manufacturing: Materials, design, technologies, and applications. Pp. 55–79.

24. Wohlers T, Campbell I, Diegel O, Kowen J, Huff R, (2019) Wohlers Report 2019. Fort Collins: Wohlers Associates, Inc. 369 p.
25. Schmidtke K, Palm F, Hawkins A, Emmelmann C, (2011) Process and mechanical properties: applicability of a scandium modified Al-alloy for laser additive manufacturing, Phys. Procedia 12: 369–374.
26. Palm F, Schmidtke K, Exceptional grain refinement in directly built up Sc-modified AlMg-alloys is promising a quantum leap in ultimate light weight design, Proc of the 9th Int Conf Trends in Welding Research, June 4th–8th 2012, Chicago, USA, 108–115.
27. Yang L, Hsu K, Baughman B, Godfrey D, Medina F, Menon M, Wiener S, (2017) Additive Manufacturing of Metals: The Technology, Materials, Design and Production. Cham: Springer. 168 p.
28. Kurzynowski T, Gruber K, Stopyra W, Kuźnicka B, Chlebus E, (2018) Correlation between process parameters, microstructure and properties of 316 L stainless steel processed by selective laser melting. Materials Science and Engineering 718: 64–73. https://doi.org/10.1016/j.msea.2018.01.103.
29. Qian M, (2015) Metal powder for additive manufacturing. Minerals, Metals & Materials Society 67(3): 536–537. https://doi.org/10.1007/s11837-015-1321-z.
30. Wei L, Ehrlich L, Poweill-Palm M, Montgomery C, Beuth J, Malen J, (2018) Thermal conductivity of metal powders for powder bed additive manufacturing, Additive Manufacturing 21: 201–208.
31. Brandt M, (2017) Laser additive manufacturing: Materials, design, technologies, and applications. Amsterdam: Elsevier. 498 p.
32. Milewski J, (2017) Additive Manufacturing of Metals: From Fundamental Technology to Rocket Nozzles, Medical Implants, and Custom Jewelry. Cham: Springer. 343 p.
33. Xjet, (2019) Technology. Internet document, Cited 16.1.2019. https://xjet3d.com/technology/.
34. Vadlamannati S, Vipin S, Soumitra B, (2019) Metal Powder Based Additive Manufacturing Technologies—Business Forecast. n 3D Printing and Additive Manufacturing Technologies. First edition. Singapore: Springer. 311 p.
35. Cabrini M, Lorenzi S, Pastore T, Pellegrini S, Ambrosio E, Calignano F, Manfredi D, Pavese M, Fino P, (2016) Effect of heat treatment on corrosion resistance of DMLS AlSi10Mg alloy. Electrochimica Acta. 206: 346–355.
36. Wei P, Wei Z, Chen Z, Du J, He Y, Li J, Zhou Y, (2017) The AlSi10Mg samples produced by selective laser melting: single track, densification, microstructure and mechanical behavior. Applied Surface Science 408: 38–50. https://doi.org/10.1016/j.apsusc.2017.02.215.
37. Bertoli U S, Guss G, Wu S, Matthews M J, Schoenung J M, (2017) In-situ characterization of laser-powder interaction and cooling rates through high-speed imaging of powder bed fusion additive manufacturing. Materials and Design 135: 385–396.

38. Mukherjee T, Wei H L, De A, DebRoy T, (2018) Heat and fluid flow in additive manufacturing—Part II: Powder bed fusion of stainless steel, and titanium, nickel and aluminum base alloys. Computational Materials Science 150: 369–380.
39. Wu A S, Brown D W, Kumar M, Gallegos G F, King W E, (2014) An experimental investigation into additive manufacturing-induced residual stresses in 316L stainless steel. Metallurgical and Materials Transactions: A 45A: 6260–6270.
40. Drummer D, Medina-Hernández M, Drexler M, Wudy K, (2015) Polymer powder production for laser melting through immiscible blends. Procedia Engineering 102: 1918–1925.
41. Concept Laser, (2019) Brochure of X LINE 2000R. Internet document, Cited 15.1.2019. https://www.concept-laser.de/fileadmin/Machine_brochures/CL_X_LINE_2000R_DS_EN_US_4_v1.pdf.
42. EOS, (2019) EOS M400–4. Internet document, Cited 11.1.2019. https://www.eos.info/systems_solutions/eos-m-400-4.
43. Renishaw, (2019) RenAM 500Q. Internet document, Cited 23.1.2019. https://www.renishaw.com/en/renam-500q%2D%2D42781.
44. SLM, (2019) SLM®800. Internet document, Cited 11.1.2019. https://slm-solutions.com/en/products/machines/slmr800/.
45. EOS, (2017) Material data sheet—EOS Maraging Steel MS1. Available: https://cdn.eos.info/1deee2b550955632/b3615b80c80a/MS-MS1-M290_Material_data_sheet_10-17_en.pdf.
46. Diegel, O, Nordin, A, Motte, D, (2019) A Practical Guide to Design for Additive Manufacturing. First edition. Singapore: Springer. 218 p.
47. Attaran, M, (2017) The rise of 3D printing: The advantages of additive manufacturing over traditional manufacturing. Business Horizons 60: 677–688.
48. Straub J, (2015) In search of technology readiness level (TRL) 10. Aerospace Science and Technology 46: 312–320.
49. Accelerating 3D manufacturing at RAPID + TCT, (2018) Manufacturing Engineering 160(4): 88–95.
50. Bagherifard S, Monti S, Zuccoli M V, Riccio M, Kondás J, Guagliano M, (2018) Cold spray deposition for additive manufacturing of freeform structural components compared to selective laser melting. Materials Science & Engineering A 721: 339–350.
51. Senvol, (2019) Senvol database of industrial AM machines and materials. Internet document, Cited 30.1.2020. http://senvol.com/5_material-results/.
52. Larimian T, Borkar T, (2019) Additive Manufacturing of In Situ Metal Matrix Composites. In: Additive Manufacturing of Emerging Materials. Cham: Springer, 1–28.
53. Leary M, (2017) Surface roughness optimization for selective laser melting (SLM): accommodating relevant and irrelevant surface. Laser additive manufacturing: Materials, design, technologies, and applications. Pp. 99–116.

Additive Manufacturing from the Point of View of Materials Research

Ville Laitinen, Mahdi Merabtene, Erica Stevens, Markus Chmielus, Jan Van Humbeeck, and Kari Ullakko

1 Introduction

Over the course of history, there have been three major industrial revolutions, each of them powered by the technological advances of the time and characterized by an increased productivity of industrial processes. Industry 1.0 incorporated the use of hydropower, steam power, and the

V. Laitinen (✉) • M. Merabtene • K. Ullakko
Material Physics Laboratory, Lappeenranta-Lahti University of Technology LUT, Lappeenranta, Finland
e-mail: ville.laitinen@lut.fi; mahdi.merabtene@student.lut.fi; kari.ullakko@lut.fi

E. Stevens • M. Chmielus
Department of Mechanical Engineering and Materials Science, University of Pittsburgh, Pittsburgh, PA, USA
e-mail: ericastevens@pitt.edu; chmielus@pitt.edu

J. Van Humbeeck
Department of Materials Engineering, KU Leuven, Leuven, Belgium
e-mail: jan.vanhumbeeck@kuleuven.be

M. Collan, K.-E. Michelsen (eds.), *Technical, Economic and Societal Effects of Manufacturing 4.0*,
https://doi.org/10.1007/978-3-030-46103-4_3

development of machine tools that enabled the mechanization of manufacturing processes; Industry 2.0 introduced mass production assembly lines that were powered by electrical energy; and Industry 3.0 introduced production automation, robots, and computer systems [1, 2]. The key aspect of the ongoing industrial revolution, Industry 4.0, relates to the cyber-physical production systems that consist of physical machines controlled and interconnected by collaborating computational elements. In fact, Industry 4.0 is strongly influenced by our ability to process data, which has phenomenally increased over the past 15 years. In parallel with Industry 4.0, there also exists the concept of Materials 4.0 (or big data materials informatics), which incorporates the tools of cyber-physical space and materials informatics to enhance the design of materials and devices with targeted functionalities in a virtual environment through computational synthesis or reverse engineering from existing knowledge on materials [3, 4]. This approach aims at a higher efficiency in synthesizing and testing novel material compositions and allows shorter lead times from conceptualization to production. However, as the concept of Materials 4.0 has been extensively reviewed in a recent article by [3], it is not discussed further in this chapter. Instead, we focus on the emerging topic of the additive manufacturing (AM) of metal-based stimuli-responsive materials and emphasize possible future directions for the additive manufacturing of metallic materials in general.

'Smart manufacturing' (later Manufacturing 4.0) is one of the primary concepts under Industry 4.0, and it can be described as an adaptable manufacturing system where production processes can adjust automatically for multiple types of products or changing conditions [1]. Manufacturing 4.0 incorporates a large group of base technologies, such as robots and other manufacturing automation, artificial intelligence, the internet of things, analytics and big data [2]. Additive manufacturing, also known as 3D printing, is without a doubt one of the key technologies empowering manufacturing under Industry 4.0. Additive manufacturing is a general term for technologies that are based on the layer-by-layer deposition of material according to a digital model of the object to be manufactured. Additive manufacturing offers many advantages, such as mass customization, reduced tooling costs, on-demand manufacturing, shorter lead times, reduced material waste, and the application-oriented optimization of geometries. In principle, additive manufacturing facilitates a greater freedom of design compared to traditional manufacturing technologies, which has opened up new ways to conduct engineering design. One of the central aspects in this development has been design for additive manufacturing (DFAM), which is a method that aims to consider additive manufacturing processes and material-related constraints in the design of components for additive manufacturing [5].

Besides freedom of design and enhanced shape complexity, another advantage of additive manufacturing relates to the materials themselves. Additive manufacturing is already today suitable for realizing complex geometries using several engineering materials, such as polymers, metals, ceramics, and composites [5–8]. Additive manufacturing has proven to be feasible for the processing of metallic materials, such as tungsten, which have been considered difficult to work with using conventional methods because of their high hardness and low ductility. In fact, for the last few years, pure tungsten has been commercially available for use in additive manufacturing systems made by EOS GmbH. Additionally, some additive manufacturing processes may introduce new options for metallic materials and enable the engineering and manufacturing of materials that are difficult or nearly impossible to synthesize using conventional methods. A good example of such materials are the so-called functionally graded materials, in which tailored properties can be obtained through a spatial gradation of chemical composition (gradient materials) and/or a 3D structure (hierarchical metamaterials). In addition, the size of these compositional or structural features can span multiple orders of magnitude. Furthermore, the introduction of new materials allows an expansion of the design space for additive manufacturing, which is interconnected with another interesting concept under Industry 4.0: the so-called 'smart materials' [9, 10].

Because materials themselves cannot be smart but can rather only exhibit certain intrinsic characteristics, the expressions 'smart materials' or 'intelligent materials' are typically (but not exclusively) used as an analogy to stimuli-responsive materials that can change their physical properties in response to external stimuli, such as a temperature change, mechanical stress, a magnetic field or an electrical current. In the scientific literature, stimuli-responsive materials are often divided into different classes based on their responses to an applied stimulus. Here, we entertain a similar approach and divide the stimuli-responsive materials into the four classes listed below.

- ***Stimuli-responsive actuator materials***—materials that produce strain in response to the applied stimuli.
- ***Stimuli-responsive energy conversion materials***—materials that exhibit an electric current, electrical resistance, magnetic field or temperature change as a primary response to the applied stimuli.

- ***Stimuli-responsive optical materials***—materials that exhibit an optical response, such as light emission or a change in optical properties, as a response to the applied stimuli.
- ***Stimuli-responsive state-changing materials***—materials that alter their physical properties, such as viscosity, in response to the applied stimuli.

Examples of stimuli-responsive materials and some of their applications are listed in Table 1, based on research by [11–88]. Applications of stimuli-responsive materials under Industry 4.0 range from small actuators, sensors, and signalization devices all the way to photovoltaic materials used in the production of electricity from sunlight. In general, stimuli-responsive materials may yield a multitude of enhanced capabilities and functionalities for many products as these allow an active response to be achieved in a product that would otherwise lack it. Some examples of applications for stimuli-responsive materials under Manufacturing 4.0 are listed below; refer to Table 1 for specific examples and references.

- Materials that can generate significant mechanical motion with almost no other components besides the material itself have a high potential for replacing traditional mechanical components, such as the gears, shafts, and pulleys that are used to **generate motion** in conventional machines. Some of these materials, such as thermally activated shape memory alloys (SMAs) or magnetic shape memory alloys (MSMAs), can still produce motion below the size threshold where mechanical components or traditional mechanisms can no longer be used, thus offering a feasible application in different types of **microelectromechanical systems**. Additionally, some of these materials, such as the shape memory alloy Ni-Ti or some of the shape memory polymers, are highly appreciated due to their biocompatibility for medical applications. Stimuli-responsive actuators can also be practical in any **soft robotics** that may be required for the handling of delicate or brittle materials or even living organisms.
- Some stimuli-responsive materials, such as magnetorheological liquids or the magnetic shape memory alloy Ni-Mn-Ga, may be practically useful in **shock absorption** and **active vibration damping,** for example in high-precision devices.
- Shape memory polymers can be used in **active disassembly** systems that are triggered at specific temperatures.
- Magnetocaloric materials can be used for high-efficiency **magnetic cooling and refrigeration** systems.

Table 1 Classification of different types of stimuli-responsive materials and examples of their applications

Class	*Group*	*Stimulus*	*Response*	*Example materials*	*Example applications*
Stimuli-responsive actuator materials	Chemomechanical polymers	Chemical reaction	Strain	Hydrogels [11]	Delivery of drugs such as insulin, artificial muscles, tissue engineering [11]
	Electroactive polymers	Electric current	Strain	Polyaniline, Nafion, polythiophenes, poly(vinyl alcohol) gel with dimethyl sulfoxide, poly(acrylonitrile) with conductive fibers [12], polypyrrole, poly(3,4-ethylene dioxythiophene) [12, 13]	Robotic applications, space applications, biomimetic applications [12], wearable sensors, prosthetic applications, haptic sensing, pulse rate monitoring, muscle movement detectors [13]
	Electrostrictive	Electric current	Strain	Lead magnesium niobite [14, 15], barium titanate [16]	Piezo-actuator applications [14, 15], actuators, transducers, energy harvesters [17]
	Magnetostrictive	Magnetic field	Strain	Metglas, Terfenol-D [18–20]	Microwave devices, sensors, transducers [18], sonar, ultrasonic cleaning [20]
	Photostrictive	Light	Strain	PLZT [21, 22]	Non-contact actuation [21], wireless remote control [21, 22]
	Shape memory alloys	Shape memory effect Temperature	Strain	Ni-Ti [23, 24], Fe-Mn-Si, Cu-Zn-Al, Cu-Al-Ni [23]	Thermal actuators, couplings [23], dental implants, artificial heart valves [23, 24], active disassembly [25, 26]
	Shape memory polymer	Shape memory effect Temperature, electric current, light	Strain	Polyurethane [27–29]	Active disassembly [30], ocular implants, vascular stents, sutures [27], self-tightening sutures [28], autochoke elements, intravenous cannula [29]
	Magnetic shape memory alloys	Shape memory effect Magnetic shape memory effect Magnetic field	Strain	Fe-Pd [31, 32], Ni-Mn-Ga [33–35], Ni-Mn-Al [36],	Micropumps [33], magnetic sensors, energy harvesters [32, 35, 37], actuators [34, 36, 38]

(*continued*)

Table 1 (continued)

Class	*Group*	*Stimulus*	*Response*	*Example materials*	*Example applications*
Stimuli-responsive energy conversion materials	Magnetoelectric	Magnetic field	Electric current	Multiferroics [39, 40]	Information storage [40], energy harvesting, resonators [42]
		Electric current	Magnetic field	Cr_2O_3 [41, 42]	optoelectronics, transducers, radioelectronics [43] microwaves [44], sensors, optical components [45, 46], spintronics and medicine [46]
	Photovoltaic	Light	Electric current	Silicon [47], gallium arsenide [48], Chalcogenide [49]	PV solar applications [47]
	Magnetocaloric	Magnetic field	Temperature change	Ni-Co-Mn-Sn [50]	Magnetic refrigeration [50]
	Pyroelectric	Temperature	Electric current	Barium strontium titanate [51]	Infrared detector [51]
	Piezoelectric	Mechanical stress	Electric current	Gallium arsenide [52], lead-zirconate-titanate [53]	Microactuators [52], biosensors [54]
	Piezoresistive	Mechanical stress	Electrical resistance	Silicon [55, 56]	Pressure sensors [55] accelerometers [56]
	Thermoelectric	Temperature	Electric current	Silicon/SiGe [57], bismuth telluride, PbTe, $SrTiO_3$ [58]	Engines, refrigeration, air conditioning [57], automotive HVAC [58]

Stimuli-responsive optical materials	Electrochromic/ Electroreflective	Electric current	Color/optical property change	Poly(indole-6-carboxylic acid) [59], polyacrylate [60], 3,4-ethylenedioxythiophene [61, 62]	Energy-saving, light control [59] supercapacitors, transistors, diodes [60], switchable mirrors [63], optical devices [61], optical recordings, forensics, smart windows [62]
	Chemochromic	Chemical reaction	Color change	Tungsten oxide [64, 65]	Hydrogen gas detectors/sensors [64, 65]
	Photochromic/ photorefractive	Light	Color change	Azobenzene, dithienylethene [66] spiropyrane [67, 68] LiNbO3:Mn [69] BaTiO3 [70]	Optical data storage, optical switching [66], drug delivery systems [67], gel glasses [68], holographic storage [69], image processing, laser resonators [70]
	Thermochromic	Temperature	Color change	Vanadium dioxide [71, 72] titanium nitride [72]	Energy saving [71, 72]
	Mechanochromic	Mechanical stress	Color change	Poly(1,4-butylene succinate) [73], polydiacetylenes, spiropyrans [74]	Healthcare [73], protective helmets, biosensors [74]
	Photoluminescent	Light	Light emission	Strontium aluminate [75, 76], zinc sulfide [76]	Warning signs, dial plates, escape routes [75], footpaths, lanes [76]
	Thermoluminescent	Temperature	Light emission	Lithium borate, beryllium oxide, lithium fluoride, calcium sulphate [77]	Medical dosimetry [77]
	Electroluminescent	Electric current	Light emission	Zinc sulfide doped with terbium or manganese [78], zinc sulfide doped with copper [79]	Emergency signage, lighting, backlights [78, 79], detection of electric/magnetic fields [80]
	Mechanoluminescent	Mechanical stress	Light emission	Zinc sulfide doped with copper, zinc sulfide doped with manganese [80]	Dynamic pressure mapping, stress sensing [80], optoelectronic devices, security papers [81]

(*continued*)

Table 1 (continued)

Class	*Group*	*Stimulus*	*Response*	*Example materials*	*Example applications*
Stimuli-responsive state changing materials	Shear-thickening	Mechanical stress	State change (fluid/solid)	Silica particles in a polyethylene glycol solution [82, 83], amphiphilic polymer, hydrophilic particles and polyethylene-oxide [82]	Adaptive stiffness, damping, smart structures [82], protective equipment, seismic isolation of buildings/bridges, shock loading applications [83]
	Electrorheological	Electric current	State change (fluid/solid)	Aluminosilicate, biopolymers (silicone oil/ hydrocarbon), oxides (surfactant) and TiO2, mineral oil, water [84, 85], zeolite and silicone oil [84]	Damping devices (shock absorbers, bearing dampers), clutches, mechanical polishers [84], haptic sensors, force feedback systems, anti-lock braking systems [85]
	Magnetorheological	Magnetic field	State change (fluid/solid)	Carbonyl iron particles and carrier liquids (silicone oils, polyesters, synthetic hydrocarbons) [86] hydrocarbon-based magneto-rheological [87]	Biomedical applications (prosthetic knees) [87], shock absorbers and dampers, brakes, clutches [86–88]

- Stimuli-responsive materials also have a high potential in different types of signalization devices, such as **displays** or **haptic** (sense of touch) technologies. In fact, haptic devices provide a unique interface between humans and machines, allowing remote distance operators to receive force feedback from the operated machines. For example, operators could receive information about the weight or resistance of lifted objects or be alerted when there is an issue with the operated machine.
- Another group of applications for stimuli-responsive materials under Industry 4.0 are different types of **sensors**, such as the ones used for failure detection and predictive maintenance in manufacturing systems. Additionally, **wearable sensors** are a prominent group of applications for many stimuli-responsive materials.

2 Additive Manufacturing of Stimuli-Responsive Materials

When it comes to stimuli-responsive materials, additive manufacturing is often referred to as 4D printing, which may refer to either the stimuli-responsive properties of the additively manufactured material in general or the ability of some of the materials (stimuli-responsive actuator materials) to change their physical shape in response to an applied stimulus. However, here we employ the term 'additive manufacturing of stimuli-responsive materials' instead of 4D printing as the usage of the former aligns better with the existing standardized terminology for additive manufacturing.

The additive manufacturing of different stimuli-responsive materials has gained significant interest in the past few years as this technology could facilitate a higher freedom of design concerning the stimuli-responsive properties of the manufactured objects. Tremendous advantages can be gained when devices can be optimized to fulfill the requirements of the intended application, instead of designing within the limits of the used manufacturing process. Thus, additive manufacturing may also accelerate the adoption of stimuli-responsive materials or expand their possible applications. Additionally, a combination of structural and stimuli-responsive materials under a single additive manufacturing process could enable the manufacturing of entire devices with integrated stimuli-responsive sections. In this case, certain functional characteristics or properties would be obtained locally in certain sections of the additively manufactured device. For example, in the case of stimuli-responsive actuator materials, the stimuli-responsive material would replace the traditional mechanisms within the manufactured device. These 'active regions' of the device could be actuated using a passive source of energy, such as a magnetic field in the case of magnetic shape memory alloys or heat in the case of thermally activated shape memory alloys. Additionally, additive manufacturing could allow a localized tailoring of properties (as in Fig. 1) within a single device,

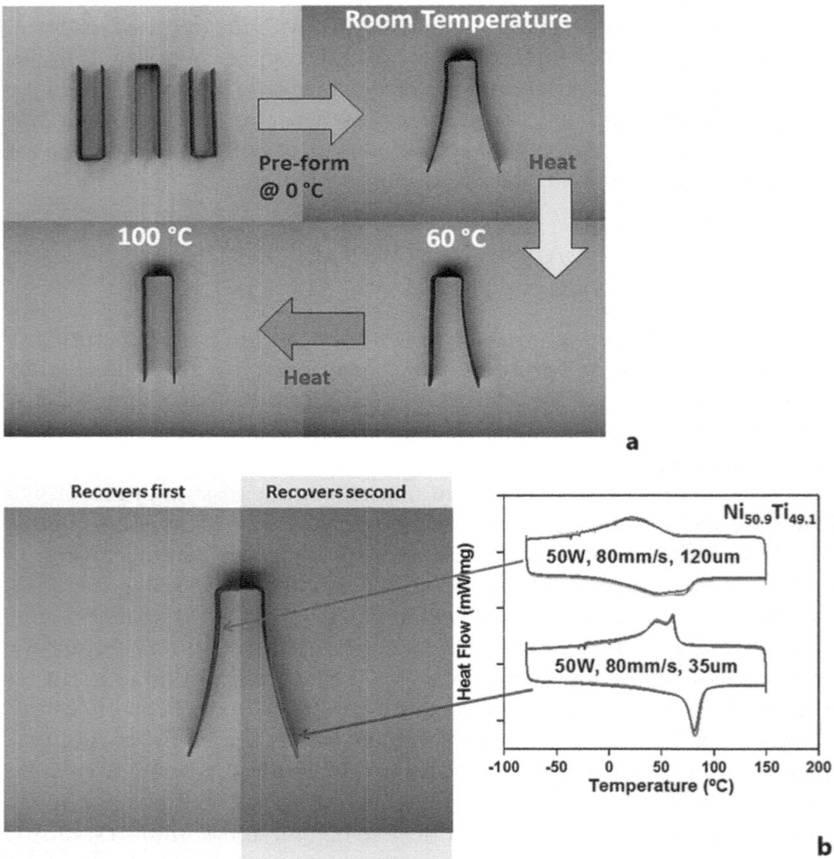

Fig. 1 (a) A location-dependent active response generated by temperature-dependent multi-stage shape recovery in a U-shaped Ni-Ti component deposited using L-PBF; (b) effect of the L-PBF process parameters on the transformation temperatures and active responses at different sections of the build. Reproduced from [151] under Creative Commons Attribution 4.0 International License

for example by inducing local differences in composition or microstructure in the processed stimuli-responsive material. Overall, these developments could facilitate the additive manufacturing of entire devices with embedded actuators or sensors, which could act as functional parts in existing systems, such as in soft robotics or pneumatics.

The majority of the published reviews on the additive manufacturing of stimuli-responsive materials have focused on shape memory polymers [152–161], while a few articles [162–167] have discussed aspects of expanding the DFAM method towards the adoption of these materials in additively manufactured components. Although some reviews have also discussed the additive manufacturing of thermally activated shape memory alloys, reviews concerning other metal-based stimuli-responsive materials, such as magnetic shape memory alloys or magnetocaloric materials, are sparse to non-existent. The popularity of polymer-based materials is expected because they are more feasible for low-cost additive manufacturing in comparison to metal-based materials, which are more difficult to manufacture additively without defects. Hence, this chapter concentrates on the additive manufacturing of thermally activated shape memory alloys, magnetic shape memory alloys, and magnetocaloric alloys. A brief overview of the state of the art in the additive manufacturing of these materials is presented in Table 2, based on the research results from [89–150]. An overview of the main additive manufacturing process categories (compared to the additive manufacturing processes in Table 2) for metal-based stimuli-responsive materials is presented below, following the definitions given in standard SFS-EN ISO/ASTM 52900:2017.

- ***Material extrusion***—"An additive manufacturing process in which material is selectively dispensed through a nozzle or orifice"; an example process for metals is 3D ink printing, whereby metal powder is dispensed in a mixture with a bonding agent.
- ***Powder bed fusion***—"An additive manufacturing process in which thermal energy selectively fuses regions of a powder bed"; the applied thermal energy can be either a laser (L-PBF) or an electron beam (E-PBF).
- ***Binder jetting***—"An additive manufacturing process in which a liquid bonding agent is selectively deposited to join powder materials".
- ***Directed energy deposition***—"An additive manufacturing process in which focused thermal energy is used to fuse materials by melting as they are being deposited"; example processes include laser-based directed energy deposition of powder material (L-DED), plasma arc deposition (PAD), and wire and arc additive manufacturing (WAAM).

Table 2 A brief overview of state of the art in the additive manufacturing of metal-based stimuli-responsive materials

Group	*Alloy*	*Process*	*Feedstock*	*Observations*
SMA	Ni-Ti	L-DED	Pre-alloyed powder	Retention and finer grain sizes proved to have a higher hardness values due to a high temperature austenite phase [89, 90]
			Mechanically produced powder	Ni-rich Ni-Ti alloys required aging treatment [91]
				Solutionized and aged Ni-Ti had a better shape memory response [92]
				Ni-rich Ni-Ti showed superelastic behavior [93]
				Shape memory effect recovery and unstable oriented/de-twinned martensite [94]
				NiTi50 showed excellent mechanical properties as compared to NiTi45 and NiTi55 [95]
			Gas atomized powder	Using low scanning speed reduced the obtained shape memory effect [96]
				Superelastic behavior [96]
				Loss of nickel in the process [97]
		L-PBF		High relative density (>97%) and hardness reported [98, 99]
				Excellent compression fatigue resistance [100] with an irreversible stain behavior [99]
				Wider hatch distance decreased relative density [101]
				Superelastic response (95%) [102–108]
				Shape memory effect [94, 105, 109, 110]
				Recovery above 5.5% [106–108, 111]
				Desired stiffness was achieved by regulating the level of porosity and stiffness reduced from 69 GPa to 20.5 GPa for 58% porosity [112]
				Loss of nickel in the process [113, 114]
				Wider hatch spacing led to a highly irrecoverable strain [114]
				Microstructure influenced the shape memory response and mechanical behavior [109, 115]
				Martensite twins were formed easily after annealing process [116]
				Heat treatment above 400 °C decreased the shape recovery and transformation strain [117]
				Tensile strain recovery was 1.5 times larger than in compression [111]

		Plasma atomized	Reversible martensitic transformation. Processing parameters, Ni evaporation and oxygen content of the processing atmosphere (oxidation) impacted martensite transformation. Solution treatment resulted in sharpened transformation peaks and undid the influence of dissolvable Ni-Ti precipitates [118].
	Micro L-PBF	Mechanically produced powder	Mechanical shape memory functionality was proven with a force range of 10–100 mN [119]
			Quality of deposited Ni-Ti was improved by increasing the scanning speed [120]
	E-PBF	PREP atomized powder	No visible shape memory and pseudoplastic effect seen [121]
			The authors did not recommend E-PBF for Ni-Ti alloys. Preheating was required [97]
			E-PBF process resulted in better tensile properties than L-DED and L-PBF [97, 122]
	PAD	Pre-alloyed powder	Linear superelasticity [123]
			Quasi-linear superelasticity with narrow hysteresis [123, 124]
	WAAM	Wire	High hardness and tensile strength [125]
Cu-Al-Ni	L-PBF	Elemental powders	High aluminum content led to dendrites and high hardness [126]
Cu-Al-Ni-Mn		Gas atomized powder	High relative density (>92%) achieved [127–129]
			Reversible martensitic transformation with the formation of β_1'-martensite [127–129]
			Large strain recovery after unloading (up to 18%) [127]
			Strong distribution of pores produced by the L-PBF sample [127, 128]
			Additional re-melting led to smaller grain size and yielded a deformability of 14% [129]
			Higher strength and improved plasticity was observed for both samples (Cu-Al-Ni-Mn and Cu-Al-Ni-Mn-Zr) [128–130]
			For the Cu-Al-Ni-Mn-Zr sample, Zr-rich phase was found to precipitate at the grain boundaries during the annealing process [130]
Cu-Al-Ni-Mn-Zr			
Cu-Al-Ni-Ti			Copper alloy with Ti addition had a high hardness of about 280 HV due to the grain refinement. The relative density exceeded 99% [131]
Fe-Mn-Al-Ni			Reversible martensitic transformation and pseudo-elastic effect [132]

(*continued*)

Table 2 (continued)

Group	*Alloy*	*Process*	*Feedstock*	*Observations*
MSMA	Ni-Mn-Ga	3D ink printing	Ink with elemental powders	Reversible martensitic transformation [133]
				Martensitic twins [134]
		Binder jetting	Mechanically produced powder	Martensitic twins and reversible martensitic transformation after post-processing [135, 136] Binder jetting produced a complex-shaped porous Ni-Mn-Ga geometry with a reversible martensitic transformation [137–139]
				Sintering of Ni-Mn-Ga powder was shown to produce net-shaped porous structures [136, 138, 139]
				Minor loss of nickel seen in sintered Ni-Mn-Ga in comparison to the initial powders [140]
		L-DED		Narrow hysteresis and increase in saturation magnetization [141]
		L-PBF	Gas atomized powder	Ni-Mn-Ga could be deposited on compositionally dissimilar materials [142, 143]
				Complex geometries [143, 144]
				High relative density of 98.3% [143]
				Martensitic twins and reversible martensitic transformation [143–146]
				Loss of Mn in process [143, 145, 146]
				Susceptibility to cracking [145]
Magnetocaloric	Ni-Co-Mn-Sn	L-DED	Mechanically produced powder	Reduction of Sn was observed at the subgrain boundaries [147]
				'Properly built' samples of Ni-Co-Mn-Sn exhibited better magnetic properties than 'overbuilt' samples [148]
				Heat-treatment promoted martensite growth and increased twinning [149]
	Ni-Mn-Cu-Ga	Binder jetting		Larger grain growth and Mn-O particles were observed at samples sintered at 1080 °C as compared 1050 °C and 1070 °C [150]

3 Additive Manufacturing of Shape Memory Alloys

Shape memory alloys are alloys that can recover a limited applied strain of less than 10% either thermally or mechanically [168]. This property finds its origin in a thermoelastic martensitic transformation in some particular alloys. This transformation is characterized by its transformation temperatures (M_s, M_f during cooling reaching the martensitic phase, A_s, A_f during heating reaching the beta phase), exhibiting a relatively small hysteresis of about 10–40 K compared to the well-known martensitic transformation in many steels exhibiting a hysteresis of several 100 K. When a strain, limited to 10%, is applied in the martensitic state, this strain can be recovered by heating above the transformation temperature into the beta phase. This is called the thermal recovery or shape memory effect. When a strain of less than 10% is applied in the beta phase, above M_s, the strain is mechanically recovered upon releasing the applied stress, which is called superelasticity. Complete thermal or mechanical recovery can only be obtained in a limited temperature window around the martensitic transformation. The thermally activated shape memory effect occurs in some Cu-based alloys and Fe-based alloys, but it is mostly associated with Ni-Ti alloys. Ni-Ti is superior compared to other shape memory alloys for many reasons, including its high ductility, high strength, and very fine grain size. These properties enable the production of very thin devices (wires with a diameter down to 25 μm). Additionally, it is biocompatible, which is why more than 80% of the products made of Ni-Ti are medically related [169]. Besides their use in medical applications, shape memory alloys can convert heat into a high force or work output, which makes these alloys useful in the actuators of stress-creating components [23, 170].

From the perspective of conventional manufacturing processes, a major problem of Ni-Ti is its poor machinability, primarily due to the strong strain hardening effect. Thus, wire and tube drawing are the most common applied forming techniques used in the production of devices such as guise wires, stents, and actuators based on springs. This sets many limitations on the shape complexity of the manufactured devices. Therefore, additive manufacturing of Ni-Ti has gained the attention of designers of medical and other devices. As shown in Table 2, laser-based processes, especially L-PBF, are the most typical approach for the additive manufacturing of Ni-Ti. The same observation applies to Cu- and Fe-based alloys, although little scientific literature is available on the additive manufacturing of these materials. In fact, the future for Cu-based SMAs does not look bright as the majority of research on additive manufacturing of SMAs concentrates on Ni-Ti. However, the additive manufacturing of Ni-Ti represents only a small fraction of the medical applications produced by metal-based additive manufacturing, and only a handful of studies on the additive manufacturing of Ni-Ti consider its stimuli-responsive properties, such as the very low stiffness (very low E-modulus), and its functional properties, such as superelasticity and the shape memory effect. However, a fair amount of research on the laser additive manufacturing (LAM) of Ni-Ti shape memory alloys has been conducted [23, 115, 171–174]. Therefore, we summarize here the most important observations of Ni-Ti deposited using L-PBF, as previously discussed by [173] and briefly overviewed in Table 2.

- Although Ni-Ti can be processed at a high density crack-free, the mechanical and functional properties of the processed material are on average inferior compared to the wrought material. However, using repetitive laser scanning in the process may allow improvement of the functional properties of deposited Ni-Ti.
- Controlling the transformation temperatures of the processed material is difficult, mainly due to the evaporation of Ni and precipitation based on impurities. Hence, the composition and transformation temperatures of the processed material are strongly dependent on the processing parameters and, therefore, the transformation temperatures of the final product are not necessarily the transformation temperatures of the initial powder.
- Additionally, the processing environment should be controlled to prevent oxygen and/or nitrogen pick-up that may lead to an increased density of impurities, which may influence the transformation temperatures and the mechanical properties of the processed material.
- The surface roughness of the final product should be considered in relation to potential wear or for the difficulties it causes in sterilization, which is required for biomedical applications.

4 Additive Manufacturing of Magnetic Shape Memory Alloys

Besides the thermally activated shape memory effect, magnetic shape memory (MSM) alloys may also exhibit a straining phenomenon when the magnetic moments of the martensitic twin variants of the alloy align with the applied magnetic field [38, 175, 176]. This straining phenomenon is called the magnetic shape memory effect. The Ni-Mn-Ga system, which is the most studied class of MSM materials, has been shown to exhibit outstanding characteristics, such as magnetic-field-induced strains (MFIS) of 12% [176], which is a hundred times larger than the magnetically induced strains obtained in competing materials. In addition, the efficiency (mechanical work output / magnetic field energy) of the MSM effect can be over 95% and its fatigue life can exceed 2×10^9 cycles [177]. Characteristic of the MSM materials is that the strain remains unchanged after the magnetic field has been switched off (the strain can be recovered by applying a magnetic field in transversal direction or by force). This results in significant energy savings in many applications, especially on-off valves, because magnetic field energy is needed only during the brief time when the shape of the MSM element is changed. Additionally, Ni-Mn-Ga can exhibit high strain accelerations of 1.6×10^6 m/s^2 [178], which is assumed to be the highest acceleration of all actuator materials. These characteristics may be beneficial in several applications, such as in robotics, biomedical applications and optics. For instance, fast actuators/sensors [34, 176], micropumps [33], and vibration energy harvesters [35] have been identified as potential applications for MSM materials. However, commercial applications of MSM materials are still limited, possibly due to the relatively young age of the technology itself compared to competing piezo ceramics or giant magnetostrictive materials.

Typically, bulk polycrystalline Ni-Mn-Ga does not exhibit limited MFIS due to grain boundary constraints that effectively block twin boundary motion in the material. However, directionally solidified (textured) polycrystalline Ni-Mn-Ga has been shown to exhibit up to 1.0% strain [179], whereas polycrystalline Ni-Mn-Ga foam has been shown to exhibit up to 8.7% recoverable strain [180]. A smaller force output and brittleness are disadvantages of foamy polycrystalline compared to more conventional single-crystalline material. From a manufacturing perspective, the use of additive manufacturing offers better freedom of design, especially compared to typical single-crystalline material. Thus, the additive manufacturing of MSM alloys aims at obtaining parts with controlled porosity while facilitating the possibility to manufacture complex geometries. Additionally, additive manufacturing places fewer limitations on the size of the manufactured object. Other advantages that may be potentially gained through additive manufacturing relate to the possibility to produce compositional gradients that allow for the tailoring of the properties for specific applications.

Compared to the additive manufacturing of Ni-Ti based thermally activated shape memory alloys, additive manufacturing of magnetic shape memory alloys is still in its infancy. All the scientific literature available at the time focuses on the additive manufacturing of Ni-Mn-Ga-based MSMAs. The most common approaches on additive manufacturing of Ni-Mn-Ga have concentrated on 3D ink printing [133, 134] and binder jetting [135–140]. However, also a few investigations into manufacturing of polycrystalline Ni-Mn-Ga using L-DED [141] and L-PBF [142–146] have recently been published. Each of the aforementioned processes have their own advantages and disadvantages concerning the manufacture of a material that exhibits MFIS. Nevertheless, a common aspect for all of the processes is the aim to obtain controlled composition, microstructure and porosity, which is essential for obtaining MFIS in polycrystalline Ni-Mn-Ga. Especially, assuring the chemical integrity of the manufactured material is also important because of the high susceptibility of the crystal structure of Ni-Mn-Ga to compositional variation and impurities.

In general, both 3D ink printing and binder jetting processes have been proven to be feasible for producing Ni-Mn-Ga with complex geometries. However, binder-based processes face a challenge regarding the control of the composition and microstructure because the consistent removal of binder elements post-processing is difficult and some oxidation and Mn evaporation may occur during the sintering process [139]. LAM processes base on melting the material, thus enabling the use of binders to be avoided. However, previous studies on the L-PBF of Ni-Mn-Ga show that some Mn is lost in the process and that this Mn loss is strongly influenced by the used process parameters [143–145]. In fact, loss of Mn during the L-PBF process is expected due to the high vapor pressure and low boiling temperature of Mn in comparison to the other elements in the alloy. Thus, control over the processing parameters and the thermal cycle that the processed material undergoes is critical for obtaining a controlled composition. This may also be an advantage, as the composition could be controlled through an adjustment of the process parameters, which could potentially allow for the adjustment of the microstructure and stimuli-responsive properties of the processed material. However, excessive over-alloying of Mn into the used powder would be required for this approach to be feasible.

The control of the porosity in the additive manufacturing of Ni-Mn-Ga has typically based on manufacturing different types of lattice structures [133, 137, 142] or foam-like materials [134, 138]. Additionally, the sintering process used in binder-based additive manufacturing processes can also be adjusted to control the density of the processed material [140]. The processed material undergoes a repetitive cycle of heating and cooling in LAM processes as the heat from melting is conducted through the prior layers of deposited material. As a result, the processed material may exhibit regions with different thermal histories, which also affects the local microstructures. This has been observed as broad ferromagnetic hysteresis and wide phase transitions in as-deposited material [141, 143]. Additionally, Ni-Mn-Ga processed by L-PBF may exhibit cracking [145]. Post-process heat-treatment is required to retain the typical ferromagnetic behavior and material properties in the deposited material [141, 143, 144]. However, laser-based processes typically produce a microstructural texture [181], which is considered beneficial for obtaining MFIS.

Although additive manufacturing shows high potential for facilitating greater design freedom for MSM based devices, so far the functional properties of the additively manufactured material are inferior compared to the conventional oriented single crystals or textured polycrystalline material. By so far, Ni-Mn-Ga processed by binder jetting [136] and L-PBF [142] have been shown to develop a magnetically induced strains up to 0.01%, which are significantly lower than the 8.7% achieved in Ni-Mn-Ga based foams [180]. In conclusion, more research on additive manufacturing of MSM alloys is required for understanding relationships between the applied process parameters and the resulting functional properties.

5 Additive Manufacturing of Magnetocaloric Materials

Some materials experience a change in entropy (Δs_T) when exposed to a magnetic field in an isothermal environment due to a phase change of either the first or second thermodynamic order [182–184]. When placed in an adiabatic environment instead, this magnetic-field-induced phase change produces a temperature change (ΔT_{ad}) in the material, leading to the common designation of this phenomenon as the magnetocaloric effect [185]. The magnetocaloric effect can be observed in both first- and second-order materials, with the order parameter of magnetization. In a first-order material, the change in magnetization is discontinuous at the transformation, whereas the change in magnetization for a second-order material is gradual and continuous over the transformation. In the case of a second-order material, this magnetization change is caused by an alignment of magnetic moments around the Curie temperature (demagnetization temperature), reducing the magnetic entropy with increasing magnetic field and causing a corresponding increase in the thermal entropy. The entropy trade-off concept remains for first-order materials, but with the addition of a magnetostructural (or magnetoelastic) phase transformation that causes the direction of the entropy change with the addition of an applied field to be less straightforward. Near the transformation temperature, an applied magnetic field will stabilize the more magnetic phase, which could be either the high-temperature or the low-temperature phase. If the high-temperature phase is stabilized, the application of a magnetic field shifts the transition to lower temperatures and leads to a decrease in the temperature of the material—called the negative (or inverse) magnetocaloric effect. If the low-temperature phase is stabilized, the application of a magnetic field shifts the transition to higher temperatures and leads to an increase in the temperature of the material—the positive magnetocaloric effect.

Recently, the magnetocaloric effect has been researched for leverage in heat pumps, particularly for cooling in applications such as solid-state-based magnetic refrigeration requiring no harmful refrigerants and in localized hypothermia therapy to treat cancer [186, 187]. For the most common application of refrigeration, any magnetocaloric effect-exhibiting material that is to be considered a viable option as a heat exchanger within a heat pump must be formed with a high surface-to-volume ratio and must allow satisfactory fluid flow [188, 189]. Thus, the following two requirements are placed upon the heat exchanger [190]:

1. Maximize the volume fraction of the magnetocaloric effect material while maintaining a large surface area.
2. Minimize the pressure drop in the fluid across the heat exchanger.

Common methods for producing heat exchanger devices from magnetocaloric effect materials are [190]: packed powder beds [191–193], parallel plates [194–197], and microchannel systems [198]. Packed powder beds, though cheap and simple, have a high pressure drop across the device due to the presence of turbulent flow. Microchannels, though inducing only a low drop in fluid pressure, have a high manufacturing cost (if they can be currently manufactured at all for the given material). Parallel plate devices are a median between the two extremes, allowing for a fluid flow that is not as turbulent as in packed powder beds and a production that is not quite as expensive as with microchannels.

With an abundance of requirements on both the feedstock material and the final magnetocaloric-effect-based heat exchanger, fabrication complications are an inescapable challenge. For example, first-order phase transition materials tend to be brittle, which limits the ability to machine them into desired geometries [199]. Difficulties with fabrication can leave promising alloys showing only a modest magnetocaloric effect after device fabrication due to changes in microstructure or atomic ordering and defects [200, 201]. In addition, first-order phase transition magnetocaloric effect materials have narrow operating temperature windows [200, 202]. For ideally efficient operation, a heat exchanger using the first-order phase transition magnetocaloric effect must have stages or a gradient of material transformation temperatures [198]. With a transformation temperature gradient, the fluid will heat (or cool) as it passes through the series of materials, at each point existing within the operating temperature for the magnetocaloric effect material that it is currently in contact with. Second-order phase transition materials are less difficult to shape and have a wider operating temperature range, but the most promising material (Gd) is a 'critical material' as it is costly, has a high environmental impact, and its use in a large number of cooling applications would lead to demand far exceeding supply [190, 199, 203].

As a manufacturing method, additive manufacturing may allow for the inclusion of designed, multi-scale porosity; complicated geometries impossible with other methods; the processing of brittle materials that cannot be machined; and gradient or layered materials with gradient or staged material transformation temperatures. This combination of benefits can grant the ability to fulfill both heat exchanger requirements with no trade-offs: a minimal pressure drop across a material that has a high surface-to-volume ratio with a maximized volume of functional material present to produce a large temperature change across a wide temperature range.

Additive manufacturing for magnetocaloric materials is in its relative infancy, although it is increasingly being recognized as a potential production avenue for magnetocaloric effect materials. In 2013, [204] used selective laser melting to create heat exchangers from $La(Fe, Co, Si)_{13}$. Meanwhile, [137, 139, 147, 148, 150] conducted experiments with Ni-Mn-based Heusler alloys fabricated using L-DED and powder bed binder jet 3D printing. L-DED, with a laser as the energy source, required a heat treatment to homogenize the microstructure before promising properties were observed [147, 148]. Binder jet printing, since it requires no heat input that would change the feedstock powder's microstructure, showed a magnetocaloric response in the as-sintered state [150] [133]. used inkjet printing to deposit a mixture containing elemental Ni, Mn, and Ga powders, then sintered them to create final lattice structures with 73–75% porosity in the micro-trusses. Published experimental studies are scarce compared to the literature on the additive manufacturing of structural metals. Nevertheless, as discussed here and in [190, 199], with the proper attention to tailoring the processing to maintain the functional properties and with measures taken to balance cost and effectiveness, additive manufacturing is a promising technology to address current manufacturing and design issues while at the same time improving the overall performance of magnetocaloric structures.

6 Future Aspects of Additive Manufacturing for Novel Metallic Materials

Besides enabling advances in freedom of design and the processing of stimuli-responsive alloys, additive manufacturing may allow the development and manufacturing of customized, application-specific materials and could thus enable the expansion of the exciting material box of different metal alloys. For example, recent developments have been made in the additive manufacturing of metal matrix composites and high-entropy alloys [8], which are favored for their outstanding mechanical properties. Additionally, significant progress has been made in engineering and manufacturing functionally graded materials, such as gradient materials or metamaterials [205–209]. A common additive manufacturing process for the fabrication of compositional gradient materials is DED, which offers unique capabilities, such as the deposition of more than one material simultaneously or the changing of the deposited material from layer to layer. A second advantage of DED is that the build process itself is not limited, compared to PBF, where deposition is only possible in successive horizontal layers. This makes the DED process suitable for depositing material on 3D substrates, such as existing parts. In fact, repairing a worn part or tool represents a typical industrial application for this process. In principle, this type of additive manufacturing process allows a precise small-scale synthesis of materials during the manufacturing process itself, thus enabling the manufacturing of materials that are difficult to synthesize on a larger scale using conventional methods. Besides potentially allowing the creation of new alloys, this also enables the application-specific tailoring of the materials of the manufacturing process itself, which could be practical for on-demand manufacturing [210]. Additionally, LAM enables the composition and microstructures to be adjusted via the process parameters, which allows the integration of information within the processed material [211]. This could be used to enhance the traceability of the used materials or processes or of the 'smart products' themselves.

7 Summary

In this chapter, we discussed how additive manufacturing could contribute to metal-based stimuli-responsive materials and material science in general. Although the future looks bright, substantial research is still required to extend the range of 'printable' materials and to achieve appropriate stimuli-responsive properties in additively manufactured metal-based materials. The complexity of the production and the material parameters create large challenges in producing dense, defect-free materials using the associated additive manufacturing processes. Indeed, specific processing conditions of metal additive manufacturing are challenging, and many material systems still suffer from cracks, unwanted porosity, high internal stresses, bad surface quality, and mechanical properties below the required levels. In many cases, this creates the need for post-processing, such as hot isostatic pressing (HIP), stress relieving, thermal treatments or polishing. However, additive manufacturing facilitates a great amount of design freedom for complex geometries and in some cases may enable the tailoring of compositional properties of the processed materials to an extent that is almost impossible to achieve using conventional manufacturing methods. Hence, additive manufacturing has a high potential for the development of novel types of stimuli-responsive devices.

References

1. Frank A, Dalenogare L, and Ayala N (2019) Industry 4.0 technologies: Implementation patterns in manufacturing companies. Int J Prod Econ 210:15–26. https://doi.org/10.1016/j.ijpe.2019.01.004.
2. Alcácer V, Cruz-Machado V (2019) Scanning the Industry 4.0: A Literature Review on Technologies for Manufacturing Systems. Eng Sci Technol Int J 22(3):899–919. https://doi.org/10.1016/j.jestch.2019.01.006.
3. Jose R, Ramakrishna S (2018) Materials 4.0: Materials big data enabled materials discovery Appl Mater Today 10:27–132. https://doi.org/10.1016/j.apmt.2017.12.015.
4. Ramakrishna S, Zhang T, Lu W, Qian Q, Low J, Yune J, Tan D, Bressan S, Sanvito S, Kalidindi S (2018) Materials informatics. J Intell Manuf 30(6):2307–2326. https://doi.org/10.1007/s10845-018-1392-0.
5. Thompson M, Moroni G, Vaneker T, Fadel G, Campbell R, Gibson I, Bernard A, Schulz J, Graf P, Ahuja B, Martina F (2016) Design for Additive Manufacturing: Trends, opportunities, considerations, and constraints. CIRP Ann 65(2):737–760. https://doi.org/10.1016/j.cirp.2016.05.004.
6. Gibson I, Rosen D, Stucker B (2015) Additive Manufacturing Technologies: 3D Printing, Rapid Prototyping, and Direct Digital Manufacturing. Springer, New York.
7. Bourell D, Kruth J, Leu M, Levy G, Rosen D, Beese A, Clare A (2017) CIRP Ann 66(2):659–681. https://doi.org/10.1016/j.cirp.2017.05.009.
8. Li N, Huang S, Zhang G, Qin R, Liu W, Xiong H, Shi G, Blackburn J (2019) Progress in additive manufacturing on new materials: A review. J Mater Sci Technol 35(2):242–269. https://doi.org/10.1016/j.jmst.2018.09.002.
9. Mittal S, Khan M, Romero D, Wuest T (2017) Smart manufacturing: Characteristics, technologies and enabling factors. Journal of Engineering Manufacture 233(5):1342–1361. https://doi.org/10.1177/0954405417736547.
10. Dilberoglu U, Gharehpapagh B, Yaman U, Dolen M (2017) The Role of Additive Manufacturing in the Era of Industry 4.0. Procedia Manuf 11:545–554. https://doi.org/10.1016/j.promfg.2017.07.148.
11. Schneider H-J, Kato K, Strongin R.M. (2007) Chemomechanical Polymers as Sensors and Actuators for Biological and Medicinal Applications. J Sens 7(8):1578–1611. https://doi.org/10.3390/s7081578.
12. Bar-Cohen Y, Zhang Q (2008) Electroactive Polymer Actuators and Sensors. MRS Bull 33(3):173–181. https://doi.org/10.1557/mrs2008.42.
13. Wang T, Farajollahi M, Choi YS, Lin I-T, Marshall JE, Thompson NM, Kar-Narayan S, Madden JDW, Smoukov SK (2016) Electroactive polymers for sensing. Interface Focus 6:20160026. https://doi.org/10.1098/rsfs.2016.0026.

14. Blackwood G, Ealey M (1993) Electrostrictive behavior in lead magnesium niobate (PMN) actuators. I. Materials perspective. Smart Mater Struct 2(2):124–133. https://doi.org/10.1088/0964-1726/2/2/008.
15. Kumar P, Singh S, Thakur O, Prakash C, Goel T (2004) Study of Lead Magnesium Niobate–Lead Titanate Ceramics for Piezo-Actuator Applications. Jpn J Appl Phys 43(4A):1501–1506. https://doi.org/10.1143/jjap.43.1501.
16. Burcsu E, Ravichandran G, Bhattacharya K (2004) Large electrostrictive actuation of barium titanate single crystals. J Mech Phys Solids, 52(4):823–846. https://doi.org/10.1016/j.jmps.2003.08.001.
17. Baraskar B, Kambale R, James A, Mahesh M, Ramana C, Kolekar Y (2017) Ferroelectric, piezoelectric and electrostrictive properties of Sn4+–modified Ba0.7Ca0.3TiO3 lead-free electroceramics. J Am Ceram Soc 100(12):5755–5765. https://doi.org/10.1111/jace.15073.
18. Nan C, Bichurin M.I, Dong S, Viehland D, Srinivasan G (2008) Multiferroic magnetoelectric composites: Historical perspective, status, and future directions. J Appl Phys 103:031101. https://doi.org/10.1063/1.2836410.
19. Olabi AG, Grunwald A (2008) Design and application of magnetostrictive materials. Mater Des 29(2):469–483. https://doi.org/10.1016/j.matdes.2006.12.016.
20. Priya S, Islam R, Dong S, Viehland D (2007) Recent advancements in magnetoelectric particulate and laminate composites. J Electroceram 19(1):147–164. https://doi.org/10.1007/s10832-007-9042-5.
21. Rahman M, Ahmed M, Nawaz M, Molina G, Rahman A (2017) Experimental Investigation of Photostrictive Materials for MEMS Application. OALib Journal 4:e3856. https://doi.org/10.4236/oalib.1103856.
22. Sun D, Tong L (2007) Modeling of wireless remote shape control for beams using nonlinear photostrictive actuators. Int J Solids Struct 44(2):672–684. https://doi.org/10.1016/j.ijsolstr.2006.05.013.
23. Mohd Jani J, Leary M, Subic A, Gibson M (2014) A review of shape memory alloy research, applications and opportunities. Mater Des 1980–2015 56:1078–1113. https://doi.org/10.1016/j.matdes.2013.11.084.
24. Mishnaevsky L, Levashov E, Valiev R, Segurado J, Sabirov I, Enikeev N, Prokoshkin S, Solov'yov A, Korotitskiy A, Gutmanas E, Gotman I, Rabkin E, Psakh'e S, Dluhoš L, Seefeldt M, Smolin A (2014) Nanostructured titanium-based materials for medical implants: Modeling and development. Mater Sci Eng R Rep 81:1–19. https://doi.org/10.1016/j.mser.2014.04.002.
25. Chiodo J, Jones N (2012) Smart materials use in active disassembly. Assembly Autom 32(1):8–24. https://doi.org/10.1108/01445151211198683.
26. Chiodo J, Jones N, Billett E (2002) WITHDRAWN: Shape memory alloy actuators for active disassembly using 'smart' materials of consumer electronic products. Mater Des https://doi.org/10.1016/s0261-3069(02)00016-x.

27. Fu Y, Huang W, Luo J, Lu H (2015) Polyurethane shape-memory polymers for biomedical applications. In: L'Hocine Y (ed) Shape Memory Polymers for Biomedical Applications. Woodhead Publishing, Cambridge, p 167–195. https://doi.org/10.1016/B978-0-85709-698-2.00009-X.
28. Lendlein A, Langer R (2002) Biodegradable, elastic shape-memory polymers for potential biomedical applications. Science, 296(5573):1673–1676. https://doi.org/10.1038/news020422-18.
29. Tobushi H, Hayashi S, Hoshio K, Ejiri Y (2008) Shape recovery and irrecoverable strain control in polyurethane shape-memory polymer. Sci Technol Adv Mat 9(1):015009. https://doi.org/10.1088/1468-6996/9/1/015009.
30. Santo L, Quadrini F, Bellisario D, Iorio L (2019) Applications of Shape-Memory Polymers, and Their Blends and Composites. In: Parameswaranpillai J, Siengchin S, George J, Jose S (ed) Shape Memory Polymers, Blends and Composites. Advanced Structured Materials vol 115. Springer, Singapore, p 311–329. https://doi.org/10.1007/978-981-13-8574-2_13.
31. Kakeshita T, Ullakko K (2002) Giant magnetostriction in ferromagnetic shape-memory alloys. MRS Bull 27(2):105–109. https://doi.org/10.1557/mrs2002.45.
32. Sánchez-Alarcos V, Recarte V, Pérez-Landazábal JI, González MA, Rodríguez-Velamazán JA (2009) Effect of Mn addition on the structural and magnetic properties of Fe-Pd ferromagnetic shape memory alloys. Acta Mater 57(14):4224–4232. https://doi.org/10.4028/www.scientific.net/MSF.635.103.
33. Ullakko K, Wendell L, Smith A, Müllner P, Hampikian G (2012) A magnetic shape memory micropump: Contact-free, and compatible with PCR and human DNA profiling. Smart Mater Struct 21:11. https://doi.org/10.1088/0964-1726/21/11/115020.
34. Smith AR, Tellinen J, Ullakko K (2014) Rapid actuation and response of Ni-Mn-Ga to magnetic-field-induced stress. Acta Mater 80:373–379. https://doi.org/10.1016/j.actamat.2014.06.054.
35. Saren A, Musiienko D, Smith A, Tellinen J, Ullakko K (2015) Modeling and design of a vibration energy harvester using the magnetic shape memory effect. Smart Mater Struct 24(9):095002. https://doi.org/10.1088/0964-1726/24/9/095002.
36. Büsgen T, Feydt J, Hassdorf R, Thienhaus S, Moske M, Boese M, Zayak A, Entel P (2004) Ab initio calculations of structure and lattice dynamics in Ni-Mn-Al shape memory alloys. Phys. Rev B 70(1):014111-1–014111-8. https://doi.org/10.1103/PhysRevB.70.014111.
37. Bruno NM, Ciocanel C, Feigenbaum HP, Waldauer A (2012) A theoretical and experimental investigation of power harvesting using the NiMnGa martensite reorientation mechanism. Smart Mater Struct 21:9 https://doi.org/10.1088/0964-1726/21/9/094018.

38. Ullakko K (1996) Magnetically controlled shape memory alloys: A new class of actuator materials. J Mater Eng Perform 5(3):405–409. https://doi.org/10.1007/BF02649344.
39. Ortega N, Kumar A, Scott J.F., Katiyar R.S. (2015) Multifunctional magnetoelectric materials for device applications. J Condens Matter Phys 27:50. https://doi.org/10.1088/0953-8984/27/50/504002.
40. Ma J, Hu J, Li Z and Nan C (2011) Recent Progress in Multiferroic Magnetoelectric Composites: from Bulk to Thin Films. Adv Mater 23(9):1062–1087. https://doi.org/10.1002/adma.201003636.
41. Meier QN, Fechner M, Nozaki T, Sahashi M, Salman Z, Prokscha T, Suter A, Schoenherr P, Lilienblum M, Borisov P, Dzyaloshinskii IE, Fiebig M, Luetkens H, Spaldin NA (2019) Search for the Magnetic Monopole at a Magnetoelectric Surface. Phys Rev X 9(1):011011. https://doi.org/10.1103/PhysRevX.9.011011.
42. Yokota T, Ichikawa K, Gomi M (2013) Epitaxial growth and magnetoelectric properties of magnetoelectric multilayers: Cr2o3/linbo3/cr2o 3 thin films. e-J Surf Sci Nanotech 11:89–92. https://doi.org/10.1380/ejssnt.2013.89.
43. Ce Wen Nan, Li M, Huang JH (2001) Calculations of giant magnetoelectric effects in ferroic composites of rare-earth-iron alloys and ferroelectric polymers. Phys. Rev B 63(14):1444151–1444159. https://doi.org/10.1103/PhysRevB.63.144415.
44. Kamenetskii EO, Sigalov M, Shavit R (2008) Magnetoelectric effect for novel microwave device applications. In: Proceedings of IEEE Convention of Electrical and Electronics Engineers in Israel, p 599. https://doi.org/10.1109/EEEI.2008.4736601.
45. Yoo I, Ahn C, Cho K (2018) 15-Mode piezoelectric composite and its application in a magnetoelectric laminate structure. J Alloys Compd 767:61–67. https://doi.org/10.1016/j.jallcom.2018.07.084.
46. Fuentes-Cobas LE, Matutes-Aquino JA, Botello-Zubiate ME, González-Vázquez A, Fuentes-Montero ME, Chateigner D (2015) Advances in Magnetoelectric Materials and Their Application, Handbook of Magnetic Materials 24:237–322. https://doi.org/10.1016/bs.hmm.2015.10.001.
47. Pizzini S (2012) Advanced Silicon Materials for Photovoltaic Applications. John Wiley & Sons, Chichester.
48. Czaban JA, Thompson DA, LaPierre RR (2009) GaAs core-shell nanowires for photovoltaic applications. Nano Lett 9(1):148–154. https://doi.org/10.1021/nl802700u.
49. Scheer R, Schock H (2011) Chalcogenide Photovoltaics: Physics, Technologies, and Thin Film Devices. Wiley-VCH, Singapore.

50. Qu YH, Cong DY, Sun XM, Nie ZH, Gui WY, Li RG, Ren Y, Wang YD (2017) Giant and reversible room-temperature magnetocaloric effect in Ti-doped Ni-Co-Mn-Sn magnetic shape memory alloys. Acta Mater 134:236–248. https://doi.org/10.1016/j.actamat.2017.06.010.
51. Cheng J, Tang J, Zhang A, Meng X, Chu J (2000) Sol-gel-derived pyroelectric barium strontium titanate thin films for infrared detector applications. Appl Phys A-Mater 71(6):667–670. https://doi.org/10.1080/00150190108016271.
52. Grudkowski TW, Montress GK, Gilden M, Black JF (1981) Integrated Circuit Compatible Surface Acoustic Wave Devices on Gallium Arsenide. IEEE Trans. Microw. Theory Tech 29(12):1348–1356. https://doi.org/10.1109/TMTT.1981.1130564.
53. Mescher M, Abe T, Brunett B, Metla H, Schlesinger TE, Reed M (1995) Piezoelectric lead-zirconate-titanate actuator films for microelectromechanical systems applications. In: Proceedings of the IEEE Micro Electro Mechanical Systems, p 261. https://doi.org/10.1109/MEMSYS.1995.472537.
54. Parton E, Bartic C, Campitelli A (2004) Micro-system platforms/new materials for biosensor applications. Solid State Technol 47(4):s1–s2.
55. Zhao Y, Fang X, Jiang Z, Zhao L (2010) An ultra-high pressure sensor based on SOI piezoresistive material. Journal of Mech Sci Technol 24(8):1655–1660. https://doi.org/10.1007/s12206-010-0515-0.
56. Tschan T, de Rooij N, Bezinge A (1992) Damping of piezoresistive silicon accelerometers. Sensor Actuat A-Phys, 32(1–3):375–379. https://doi.org/10.1016/0924-4247(92)80015-U.
57. Jovanovic V, Ghamaty S, Bass JC (2012) New thermoelectric materials and applications. In: Proceedings of InterSociety Conference on Thermal and Thermomechanical Phenomena in Electronic Systems, p 1159. https://doi.org/10.1109/ITHERM.2012.6231554.
58. Yang J, Stabler FR (2009) Automotive applications of thermoelectric materials. J Electron Mater 38(7):1245–1251. https://doi.org/10.1007/s11664-009-0680-z.
59. Kuo C, Wu T, Fan S (2018) Applications of Poly(indole-6-carboxylic acid-co-2,2′-bithiophene) Films in High-Contrast Electrochromic Devices. Coatings 8(3):102. https://doi.org/10.3390/coatings8030102.
60. Lin K, Zhang S, Liu H, Zhao Y, Wang Z, Xu J (2015) Effects on the electrochemical and electrochromic properties of 3 linked polythiophene derivative by the introduction of polyacrylate. Int J Electrochem Sci 10(9):7720–7731.
61. Fabre-Francke I, Aubert P, Alfonsi S, Vidal F, Sauques L, Chevrot C (2012) Electropolymerization of 3,4-ethylenedioxythiophene within an insulating nitrile butadiene rubber network: Application to electroreflective surfaces and devices. Sol Energy Mater Sol Cells 99:109–115. https://doi.org/10.1016/j.solmat.2011.07.004.

62. Granqvist C (2014) Electrochromics for smart windows: Oxide-based thin films and devices. Thin Solid Films 564:1–38. https://doi.org/10.1016/j.tsf.2014.02.002.
63. Tajima K, Yamada Y, Bao S, Okada M, Yoshimura K (2008) Flexible all-solid-state switchable mirror on plastic sheet. Appl Phys Lett 92(4):041912. https://doi.org/10.1063/1.2839298.
64. Smith II RD, Benson DK, Maroef I, Olson DL, Wildeman TR (2001) The determination of hydrogen distribution in high-strength steel weldments. Part 2: Opto-electronic diffusible hydrogen sensor. Weld J 80(5):122s–125s.
65. Benson, D.K., Tracy, C.E., Hishmeh, G.A., Ciszek, P.E., Lee, S. & Haberman, D.P. (1999). In: Proceedings of SPIE—The International Society for Optical Engineering 3535, p 185–202. https://doi.org/10.1117/12.337465.
66. Shirota Y, Utsumi H, Ujike T, Yoshikawa S, Moriwaki K, Nagahama D, Nakano H (2003) Photochromic amorphous molecular materials and their applications. Opt Mater 21(1–3):249–254. https://doi.org/10.1016/S0925-3467(02)00145-3.
67. Tangso KJ, Fong W-K, Darwish T, Kirby N, Boyd BJ, Hanley TL (2013) Novel spiropyran amphiphiles and their application as light-responsive liquid crystalline components. J Phys Chem B 117(35):10203–10210. https://doi.org/10.1021/jp403840m.
68. Levy D (1997) Recent applications of photochromic sol-gel materials. Mol Cryst Liq Cryst Sci Tech 297–298:31–39. https://doi.org/10.1080/10587259708036100.
69. Yang Y, Psaltis D, Luennemann M, Berben D, Hartwig U, Buse K (2003) Photorefractive properties of lithium niobate crystals doped with manganese. J Opt Soc Am B 20(7):1491. https://doi.org/10.1364/josab.20.001491.
70. Ducharme S, Feinberg J (1986) Altering the photorefractive properties of BaTiO_3 by reduction and oxidation at 650°C. J Opt Soc Am B 3(2):283. https://doi.org/10.1364/josab.3.000283.
71. Li Y, Ji S, Gao Y, Luo H, Kanehira M (2013) Core-shell VO2 @TiO2 nanorods that combine thermochromic and photocatalytic properties for application as energy-saving smart coatings. Sci Rep 3:1370 https://doi.org/10.1038/srep01370.
72. Hao Q, Li W, Xu H, Wang J, Yin Y, Wang H, Ma L, Ma F, Jiang X, Schmidt OG, Chu PK (2018) VO 2 /TiN Plasmonic Thermochromic Smart Coatings for Room-Temperature Applications. Adv Mater 30(10):1705421. https://doi.org/10.1002/adma.201705421.
73. Pucci A, Ruggeri G (2011) Mechanochromic polymer blends. J Mater Chem 21(23):8282. https://doi.org/10.1039/C0JM03653F.
74. Jiang Y (2014) An outlook review: Mechanochromic materials and their potential for biological and healthcare applications. Mater Sci Eng: C 45:682–689. https://doi.org/10.1016/j.msec.2014.08.027.

75. Gao F, Xiong Z, Xue H, Liu Y (2009) Improved performance of strontium aluminate luminous coating on the ceramic surface. J Phys 152:012082. https://doi.org/10.1088/1742-6596/152/1/012082.
76. Giuliani F, Autelitano F (2014) Photoluminescent road surface dressing: A first laboratory experimental investigation. Mater Tech 102(6–7):603. https://doi.org/10.1051/mattech/2014030.
77. Azorin J (2014) Preparation methods of thermoluminescent materials for dosimetric applications: An overview. Appl Radiat Isot 83:187–191. https://doi.org/10.1016/j.apradiso.2013.04.031.
78. Tang W, Cameron DC (1996) Electroluminescent zinc sulphide devices produced by sol-gel processing. Thin Solid Films 280(1–2):221–226. https://doi.org/10.1016/0040-6090(95)08198-4.
79. Smet P, Moreels I, Hens Z, Poelman D (2010) Luminescence in Sulfides: A Rich History and a Bright Future. Mater 3(4):2834–2883. https://doi.org/10.3390/ma3042834.
80. Feng A, Smet P (2018) A Review of Mechanoluminescence in Inorganic Solids: Compounds, Mechanisms, Models and Applications. Mater 11(4):484. https://doi.org/10.3390/ma11040484.
81. Di B, Chen Y (2018) Recent progress in organic mechanoluminescent materials. Chin Chem Lett 29(2):245–251. https://doi.org/10.1016/j.cclet.2017.08.043.
82. Ding J, Li W, SZ Shen (2011) Research and Applications of Shear Thickening Fluids. Recent Pat Mater Sci 4(1):43–49. https://doi.org/10.2174/1874464811104010043.
83. Gürgen S, Kushan M, Li W (2017) Shear thickening fluids in protective applications: A review. Prog Polym Sci 75:48–72. https://doi.org/10.1016/j.progpolymsci.2017.07.003.
84. Hao T (2002) Electrorheological suspensions. Adv Colloid Interface Sci 97(1–3):1–35. https://doi.org/10.1016/S0001-8686(01)00045-8.
85. Agafonov, A. and Zakharov, A. (2010) Electrorheological fluids. Russ J Gen Chem 80:567–575. https://doi.org/10.1134/S1070363210030382.
86. de Vicente J, Klingenberg D, Hidalgo-Alvarez R (2011) Magnetorheological fluids: a review. Soft Matter 7(8):3701. https://doi.org/10.1039/C0SM01221A.
87. Spaggiari A (2012) Properties and applications of Magnetorheological fluids. Frattura ed Integrità Strutturale 7(23):48–61. https://doi.org/10.3221/IGF-ESIS.23.06
88. Kostamo E, Kostamo J, Kajaste J, Pietola M (2012) Magnetorheological valve in servo applications. J Intel Mater Syst Str 23(9):1001–1010. https://doi.org/10.1177/1045389X12436732.

89. Krishna B, Bose S, Bandyopadhyay A (2007) Laser Processing of Net-Shape NiTi Shape Memory Alloy. Metallurgical and Materials Transactions A 38(5):1096–1103. https://doi.org/10.1007/s11661-007-9127-4.
90. Marattukalam J, Singh A, Datta S, Das M, Balla V, Bontha S, Kalpathy S (2015) Microstructure and corrosion behavior of laser processed NiTi alloy. Mater Sci Eng C 57:309–313. https://doi.org/10.1016/j.msec.2015.07.067.
91. Hamilton R, Palmer T, Bimber B (2015) Spatial characterization of the thermal-induced phase transformation throughout as-deposited additive manufactured NiTi bulk builds. Scr Mater 101:56–59. https://doi.org/10.1016/j.scriptamat.2015.01.018.
92. Hamilton R, Bimber B, Palmer T (2018) Correlating microstructure and superelasticity of directed energy deposition additive manufactured Ni-rich NiTi alloys. J Alloy Compd 739:712–722. https://doi.org/10.1016/j.jallcom.2017.12.270.
93. Bimber B, Hamilton R, Keist J, Palmer T (2016) Anisotropic microstructure and superelasticity of additive manufactured NiTi alloy bulk builds using laser directed energy deposition. Mater Sci Eng A 674:125–134. https://doi.org/10.1016/j.msea.2016.07.059.
94. Hamilton R, Bimber B, Taheri AM, Elahinia M (2017) Multi-scale shape memory effect recovery in NiTi alloys additive manufactured by selective laser melting and laser directed energy deposition. J Mater Process Technol 250:55–64. https://doi.org/10.1016/j.jmatprotec.2017.06.027.
95. Shiva S, Palani I, Mishra S, Paul C, Kukreja L (2015) Investigations on the influence of composition in the development of Ni–Ti shape memory alloy using laser based additive manufacturing. Opt Laser Technol 69:44–51. https://doi.org/10.1016/j.optlastec.2014.12.014.
96. Baran A, Polanski M (2018) Microstructure and properties of LENS (laser engineered net shaping) manufactured Ni-Ti shape memory alloy. J Alloy Compd 750:863–870. https://doi.org/10.1016/j.jallcom.2018.03.400.
97. Wang C, Tan X, Du Z, Chandra S, Sun Z, Lim C, Tor S, Lim C, Wong C (2019) Additive manufacturing of NiTi shape memory alloys using premixed powders. J Mater Process Technol 271:152–161. https://doi.org/10.1016/j.jmatprotec.2019.03.025.
98. Shishkovsky I, Yadroitsev I, Smurov I (2012) Direct Selective Laser Melting of Nitinol Powder. Phys Procedia 39:447–454. https://doi.org/10.1016/j.phpro.2012.10.060.
99. Walker J, Haberland C, Taheri AM, Karaca H, Dean D, Elahinia M (2016) Process development and characterization of additively manufactured nickel–titanium shape memory parts. J Intel Mat Syst Str 27(19):2653–2660. https://doi.org/10.1177/1045389X16635848

100. Speirs M, Van Hooreweder B, Van Humbeeck J, Kruth J (2017) Fatigue behaviour of NiTi shape memory alloy scaffolds produced by SLM, a unit cell design comparison. J Mech Behav Biomed Mater 70:53–59. https://doi.org/10.1016/j.jmbbm.2017.01.016.
101. Walker J, Elahinia M, Haberland C (2013) An Investigation of Process Parameters on Selective Laser Melting of Nitinol. In: Proceedings of the ASME 2013 Conference on Smart Materials, Adaptive Structures and Intelligent Systems Vol 1, V001T01A007. https://doi.org/10.1115/SMASIS2013-3074.
102. Bormann T, Schumacher R, Müller B, Mertmann M, de Wild M (2012) Tailoring Selective Laser Melting Process Parameters for NiTi Implants. J Mater Eng Perform 21(12):2519–2524. https://doi.org/10.1007/s11665-012-0318-9.
103. Jahadakbar A, Shayesteh MN, Amerinatanzi A, Dean D, Karaca H, Elahinia M (2016) Finite Element Simulation and Additive Manufacturing of Stiffness-Matched NiTi Fixation Hardware for Mandibular Reconstruction Surgery. Bioengineering 3(4):36. https://doi.org/10.3390/bioengineering3040036.
104. Saedi S, Turabi A, Taheri AM, Haberland C, Karaca H, Elahinia M (2016) The influence of heat treatment on the thermomechanical response of Ni-rich NiTi alloys manufactured by selective laser melting. J Alloy Compd 677:204–210. https://doi.org/10.1016/j.jallcom.2016.03.161.
105. Saedi S, Turabi A, Andani M, Haberland C, Elahinia M, Karaca H (2016) Thermomechanical characterization of Ni-rich NiTi fabricated by selective laser melting. Smart Mater Struct 25(3):035005. https://doi.org/10.1088/0964-1726/25/3/035005.
106. Saedi S, Turabi A, Andani M, Moghaddam N, Elahinia M, Karaca H (2017) Texture, aging, and superelasticity of selective laser melting fabricated Ni-rich NiTi alloys. Mater Sci Eng A 686:1–10. https://doi.org/10.1016/j.msea.2017.01.008.
107. Saedi S, Shayesteh AN, Amerinatanzi A, Elahinia M, Karaca H (2018) On the effects of selective laser melting process parameters on microstructure and thermomechanical response of Ni-rich NiTi. Acta Mater 144:552–560. https://doi.org/10.1016/j.actamat.2017.10.072.
108. Shayesteh MN, Saedi S, Amerinatanzi A, Hinojos A, Ramazani A, Kundin J, Mills M, Karaca H, Elahinia M (2019) Achieving superelasticity in additively manufactured NiTi in compression without post-process heat treatment. Sci Rep 9(1):41. https://doi.org/10.1038/s41598-018-36641-4.
109. Li S, Hassanin H, Attallah M, Adkins N, Essa K (2016) The development of TiNi-based negative Poisson's ratio structure using selective laser melting. Acta Mater 105:75–83. https://doi.org/10.1016/j.actamat.2015.12.017.

110. Taheri AM, Saedi S, Turabi A, Karamooz M, Haberland C, Karaca H, Elahinia M (2017) Mechanical and shape memory properties of porous Ni 50.1 Ti 49.9 alloys manufactured by selective laser melting. J Mech Behav Biomed 68:224–231. https://doi.org/10.1016/j.jmbbm.2017.01.047.
111. Shayesteh MN, Saghaian S, Amerinatanzi A, Ibrahim H, Li P, Toker G, Karaca H, Elahinia M (2018) Anisotropic tensile and actuation properties of NiTi fabricated with selective laser melting. Mat Sci Eng A 724:220–230. https://doi.org/10.1016/j.msea.2018.03.072.
112. Taheri AM, Haberland C, Walker J, Karamooz M, Sadi TA, Saedi S, Rahmanian R, Karaca H, Dean D, Kadkhodaei M, Elahinia M (2016) Achieving biocompatible stiffness in NiTi through additive manufacturing. J Intel Mat Syst Str 27(19):2661–2671. https://doi.org/10.1177/1045389X16641199.
113. Elahinia M, Shayesteh MN, Amerinatanzi A, Saedi S, Toker G, Karaca H, Bigelow G, Benafan O (2018) Additive manufacturing of NiTiHf high temperature shape memory alloy. Scr Mater 145:90–94. https://doi.org/10.1016/j.scriptamat.2017.10.016
114. Sam J, Franco B, Ma J, Karaman I, Elwany A, Mabe J (2018) Tensile actuation response of additively manufactured nickel-titanium shape memory alloys. Scr Mater 146:164–168. https://doi.org/10.1016/j.scriptamat.2017.11.013.
115. Dadbakhsh S, Speirs M, Van Humbeeck J, Kruth J (2016) Laser additive manufacturing of bulk and porous shape-memory NiTi alloys: From processes to potential biomedical applications. MRS Bull 41(10):765–774. https://doi.org/10.1557/mrs.2016.209.
116. Dadbakhsh S, Vrancken B, Kruth J, Luyten J, Van Humbeeck J (2016) Texture and anisotropy in selective laser melting of NiTi alloy. Mat Sci Eng A 650:225–232. https://doi.org/10.1016/j.msea.2015.10.032.
117. Khoo Z, An J, Chua C, Shen Y, Kuo C, Liu Y (2018) Effect of Heat Treatment on Repetitively Scanned SLM NiTi Shape Memory Alloy. Materials 12(1):77. https://doi.org/10.3390/ma12010077.
118. Speirs M, Wang X, Van BS, Ahadi A, Dadbakhsh S, Kruth J, Van Humbeeck J (2016) On the Transformation Behavior of NiTi Shape-Memory Alloy Produced by SLM. Shap. Mem. Superelasticity 2(4):310–316. https://doi.org/10.1007/s40830-016-0083-y.
119. Hagemann R, Noelke C, Rau T, Kaierle S, Overmeyer L, Wesling V, Wolkers W (2015) Design, processing, and characterization of nickel titanium microactuators for medical implants. J Laser Appl 27(S2):S29203. https://doi.org/10.2351/1.4906381.
120. Khademzadeh S, Parvin N, Bariani P, Mazzucato F (2015) Effects of micro laser sintering process parameters on quality of nickel-titanium single tracks and thin walls. Met and Mater Int 21(6):1081–1090. https://doi.org/10.1007/s12540-015-5304-y.

121. Hayat M, Chen G, Liu N, Khan S, Tang H, Cao P (2018) Physical and Tensile Properties of NiTi Alloy by Selective Electron Beam Melting. Key Eng Mater 770:148–154. https://doi.org/10.4028/www.scientific.net/KEM.770.148.
122. Zhou Q, Hayat M, Chen G, Cai S, Qu X, Tang H, Cao P (2019) Selective electron beam melting of NiTi: Microstructure, phase transformation and mechanical properties. Mat Sci Eng A 744:290–298. https://doi.org/10.1016/j.msea.2018.12.023.
123. Lu B, Cui X, Liu E, Feng X, Dong M, Li Y, Wang H, Jin G (2018) Influence of microstructure on phase transformation behavior and mechanical properties of plasma arc deposited shape memory alloy. Mat Sci Eng A 736:130–136. https://doi.org/10.1016/j.msea.2018.08.098.
124. Lu B, Cui X, Feng X, Dong M, Li Y, Cai Z, Wang H, Jin G (2018) Direct rapid prototyping of shape memory alloy with linear superelasticity via plasma arc deposition. Vacuum 157:65–68. https://doi.org/10.1016/j.vacuum.2018.08.028.
125. Wang J, Pan Z, Yang G, Han J, Chen X, Li H (2019) Location dependence of microstructure, phase transformation temperature and mechanical properties on Ni-rich NiTi alloy fabricated by wire arc additive manufacturing. Mat Sci Eng A 749:218–222. https://doi.org/10.1016/j.msea.2019.02.029.
126. Reyes DG, Walczak M, Ramos ME, Ramos-Grez J (2017) Towards direct metal laser fabrication of Cu-based shape memory alloys. Rapid Prototyp J 23(2):329–336. https://doi.org/10.1108/RPJ-02-2016-0017.
127. Gargarella P, Kiminami C, Mazzer E, Cava R, Basilio L, Bolfarini C, Botta W, Eckert J, Gustmann T, Pauly S (2015) Phase Formation, Thermal Stability and Mechanical Properties of a Cu-Al-Ni-Mn Shape Memory Alloy Prepared by Selective Laser Melting. Mater Res 18(2):35–38. https://doi.org/10.1590/1516-1439.338914.
128. Gustmann T, Neves A, Kühn U, Gargarella P, Kiminami C, Bolfarini C, Eckert J, Pauly S (2016) Influence of processing parameters on the fabrication of a Cu-Al-Ni-Mn shape-memory alloy by selective laser melting. Addit Manuf 11:23–31. https://doi.org/10.1016/j.addma.2016.04.003
129. Gustmann T, Schwab H, Kühn U, Pauly S (2018) Selective laser remelting of an additively manufactured Cu-Al-Ni-Mn shape-memory alloy. Mater Des 153:129–138. https://doi.org/10.1016/j.matdes.2018.05.010.
130. Gustmann T, dos Santos J, Gargarella P, Kühn U, Van Humbeeck J, Pauly S (2016) Properties of Cu-Based Shape-Memory Alloys Prepared by Selective Laser Melting. Shap. Mem. Superelasticity 3(1):24–36. https://doi.org/10.1007/s40830-016-0088-6.
131. Tian J, Zhu W, Wei Q, Wen S, Li S, Song B, Shi Y (2019) Process optimization, microstructures and mechanical properties of a Cu-based shape memory alloy fabricated by selective laser melting. J Alloy Compd 785:754–764. https://doi.org/10.1016/j.jallcom.2019.01.153.

132. Niendorf T, Brenne F, Krooß P, Vollmer M, Günther J, Schwarze D, Biermann H (2016) Microstructural Evolution and Functional Properties of Fe-Mn-Al-Ni Shape Memory Alloy Processed by Selective Laser Melting. Metall Mater Trans A 47(6):2569–2573. https://doi.org/10.1007/s11661-016-3412-z.
133. Taylor S, Shah R, Dunand D (2018) Ni-Mn-Ga micro-trusses via sintering of 3D-printed inks containing elemental powders. Acta Mater 143:20–29. https://doi.org/10.1016/j.actamat.2017.10.002.
134. Taylor S, Shah R Dunand D (2019) Microstructure and porosity evolution during sintering of Ni-Mn-Ga wires printed from inks containing elemental powders. Intermetallics, 104:113–123. https://doi.org/10.1016/j.intermet.2018.10.024.
135. Caputo M, Solomon C, Nguyen P, Berkowitz A (2016) Electron Microscopy Investigation of Binder Saturation and Microstructural Defects in Functional Parts Made by Additive Manufacturing. Microsc Microanal 22(S3):1770–1771. https://doi.org/10.1017/S1431927616009697.
136. Caputo M, Berkowitz A, Armstrong A, Müllner P, Solomon C (2018) 4D printing of net shape parts made from Ni-Mn-Ga magnetic shape-memory alloys. Addit Manuf 21:579–588. https://doi.org/10.1016/j.addma.2018.03.028.
137. Caputo M, Solomon C (2017) A facile method for producing porous parts with complex geometries from ferromagnetic Ni-Mn-Ga shape memory alloys. Mater Lett 200:87–89. https://doi.org/10.1016/j.matlet.2017.04.112.
138. Mostafaei A, Kimes K, Stevens E, Toman J, Krimer Y, Ullakko K, Chmielus M (2017) Microstructural evolution and magnetic properties of binder jet additive manufactured Ni-Mn-Ga magnetic shape memory alloy foam. Acta Mater 131:482–490. https://doi.org/10.1016/j.actamat.2017.04.010.
139. Mostafaei A, Rodriguez De Vecchis P, Stevens E, Chmielus M (2018) Sintering regimes and resulting microstructure and properties of binder jet 3D printed Ni-Mn-Ga magnetic shape memory alloys. Acta Mater 154:355–364. https://doi.org/10.1016/j.actamat.2018.05.047.
140. Caputo M, Solomon C (2017) Microstructure and Chemical Composition Analysis of Additive Manufactured Ni-Mn-Ga Parts Sintered in Different Conditions. Microsc Microanal 23(S1):2078–2079. https://doi.org/10.1017/S1431927617011059.
141. Toman J, Müllner P, Chmielus M (2018) Properties of as-deposited and heat-treated Ni-Mn-Ga magnetic shape memory alloy processed by directed energy deposition. J Alloy Compd 752:455–463. https://doi.org/10.1016/j.jallcom.2018.04.059.
142. Ullakko K, Laitinen V, Saren A, Sozinov A, Musiienko D, Chmielus M, Salminen A (2018) Ni-Mn-Ga actuating elements manufactured using 3D

printing. Paper presented at the 11th European Symposium on Martensitic Transformations, Metz, 27–31 August 2018.
143. Laitinen V, Sozinov A, Saren A, Salminen A, Ullakko K (2019) Laser powder bed fusion of Ni-Mn-Ga magnetic shape memory alloy. Addit Manuf 30C: 100891. https://doi.org/10.1016/j.addma.2019.100891.
144. Nilsén F, Ituarte I, Salmi M, Partanen J, Hannula S (2019) Effect of process parameters on non-modulated Ni-Mn-Ga alloy manufactured using powder bed fusion. Addit Manuf 28:464–474. https://doi.org/10.1016/j.addma.2019.05.029.
145. Laitinen V, Salminen A, Ullakko K (2019) First investigation on processing parameters for laser powder bed fusion of Ni-Mn-Ga magnetic shape memory alloy. J Laser Appl 31(2):022303. https://doi.org/10.2351/1.5096108.
146. Laitinen V, Sozinov A, Saren A, Ullakko K (2019) Laser based 4D printing of Ni-Mn-Ga MSM alloy. In: Book of abstracts of 6th International Conference on Ferromagnetic Shape Memory Alloys p 156-157. ISBN 978-80-905962-9-0.
147. Stevens E, Toman J, Kimes K, Chernenko V, Wojcik A, Maziarz W, Chmielus M (2016) Microstructural Evaluation of Magnetocaloric Ni-Co-Mn-Sn Produced by Directed Energy Deposition. Microsc Microanal 22(S3):1774–1775. https://doi.org/10.1017/s1431927616009715.
148. Stevens E, Kimes K, Chernenko V, Wojcik A, Maziarz W, Toman J, Chmielus M (2017). Characterization of Direct Laser Deposited Magnetocaloric Ni-Co-Mn-Sn. In: Proceedings of Materials Science and Technology, p 430–432. https://doi.org/10.7449/2017mst/2017/mst_2017_430_432.
149. Stevens E, Kimes K, Chernenko V, Lazpita P, Wojcik A, Maziarz W, Chmielus M (2018) Direct Laser Deposition and Homogenization of Ni-Co-Mn-Sn Magnetocaloric Material. Microsc Microanal 24(S1):956–957. https://doi.org/10.1017/s1431927618005275.
150. Stevens E, Salazar D, Kimes K, de Vecchis R, Chernenko V, Chmielus M (2019) Additive Manufacturing of Ni-Mn-Cu-Ga: Influence of Sintering Temperature on Magnetocaloric Effect and Microstructure. Microsc Microanal 25(S2):2578–2579. https://doi.org/10.1017/S143192761901362X.
151. Ma J, Franco B, Tapia G, Karayagiz K, Johnson L, Liu J, Arroyave R, Karaman I, Elwany A (2017) Spatial Control of Functional Response in 4D-Printed Active Metallic Structures. Sci Rep 7(1):46707. https://doi.org/10.1038/srep46707.
152. Khoo Z, Teoh J, Liu Y, Chua C, Yang S, An J, Leong K, Yeong W (2015) 3D printing of smart materials: A review on recent progresses in 4D printing. Virtual Phys Prototy 10(3):103–122. https://doi.org/10.1080/17452759.2015.1097054.
153. Leist S, Zhou J (2016) Current status of 4D printing technology and the potential of light-reactive smart materials as 4D printable materials. Virtual

Phys Prototy 11(4):249–262. https://doi.org/10.1080/17452759.2016.1198630
154. Lee J, An J, Chua C (2017) Fundamentals and applications of 3D printing for novel materials. Appl Mater Today 7:120–133. https://doi.org/10.1016/j.apmt.2017.02.004.
155. Lee A, An J, Chua C (2017) Two-Way 4D Printing: A Review on the Reversibility of 3D-Printed Shape Memory Materials. Eng 3(5):663–674. https://doi.org/10.1016/J.ENG.2017.05.014.
156. Mitchell A, Lafont U, Hołyńska M, Semprimoschnig C (2018) Additive manufacturing — A review of 4D printing and future applications. Addit Manuf 24:606–626. https://doi.org/10.1016/j.addma.2018.10.038
157. Momeni FM. Mehdi Hassani.N S, Liu X, Ni J (2017) A review of 4D printing. Mater Des 122:42–79. https://doi.org/10.1016/j.matdes.2017.02.068.
158. Boydston A, Cao B, Nelson A, Ono R, Saha A, Schwartz J, Thrasher C (2018) Additive manufacturing with stimuli-responsive materials. J Mater Chem A 6(42):20621–20645. https://doi.org/10.1039/C8TA07716A.
159. Chang J, He J, Mao M, Zhou W, Lei Q, Li X, Li D, Chua C, Zhao X (2018) Advanced Material Strategies for Next-Generation Additive Manufacturing. Materials 11(1):166. https://doi.org/10.3390/ma11010166.
160. Kuang X, Roach D, Wu J, Hamel C, Ding Z, Wang T, Dunn M, Qi H (2018) Advances in 4D Printing: Materials and Applications. Adv Funct Mater 29(2):1805290. https://doi.org/10.1002/adfm.201805290.
161. Zhang Z, Demir K, Gu G (2019) Developments in 4D-printing: a review on current smart materials, technologies, and applications. Inter J Smart Nano Mater 10:205–224. https://doi.org/10.1080/19475411.2019.1591541.
162. Li X, Shang J, Wang Z (2017) Intelligent materials: a review of applications in 4D printing. Assembly Autom 37(2):170–185. https://doi.org/10.1108/AA-11-2015-093.
163. Oliveira J, Correia V, Castro H, Martins P, Lanceros-Mendez S (2018) Polymer-based smart materials by printing technologies: Improving application and integration. Addit Manuf 21:269–283. https://doi.org/10.1016/j.addma.2018.03.012.
164. Sossou G, Demoly F, Belkebir H, Qi H, Gomes S, Montavon G (2019) Design for 4D printing: A voxel-based modeling and simulation of smart materials. Mater Des 175:107798. https://doi.org/10.1016/j.matdes.2019.107798.
165. Zhang Z, Demir K, Gu G (2019) Developments in 4D-printing: a review on current smart materials, technologies, and applications. International Journal of Smart and Nano Materials 10(3):205–224. https://doi.org/10.1080/19475411.2019.1591541.

166. Gardan J (2018) Smart materials in additive manufacturing: state of the art and trends. Virtual Phys Prototyp 14(1):1–18. https://doi.org/10.1080/17452759.2018.1518016.
167. Shafranek R, Millik S, Smith P, Lee C, Boydston A, Nelson A (2019) Stimuli-responsive materials in additive manufacturing. Prog Polym Sci 93:36–67. https://doi.org/10.1016/j.progpolymsci.2019.03.002.
168. Yamauchi K, Ohkata I, Tsuchiya K, Miyazaki S (2011) Shape memory and superelastic alloys: applications and technologies. Woodhead Publishing, Oxford.
169. Duerig T (2012) Shape Memory Alloys. In: ASM Handbook, Volume 23—Materials for medical devices. ASM International, Ohio, p 237–250.
170. Van Humbeeck J (1999) Non-medical applications of shape memory alloys. Mat Sci Eng A 273-275:134–148. https://doi.org/10.1016/S0921-5093(99)00293-2.
171. Elahinia M, Hashemi M, Tabesh M, Bhaduri S (2012) Manufacturing and processing of NiTi implants: A review. Prog Mater Sci 57(5):911–946. https://doi.org/10.1016/j.pmatsci.2011.11.001.
172. Elahinia M, Shayesteh MN, Taheri AM, Amerinatanzi A, Bimber B, Hamilton R (2016) Fabrication of NiTi through additive manufacturing: A review. Prog Mater Sci 83:630–663. https://doi.org/10.1016/j.pmatsci.2016.08.001.
173. Van Humbeeck J (2018) Additive Manufacturing of Shape Memory Alloys. Shap. Mem. Superelasticity 4(2):309–312. https://doi.org/10.1007/s40830-018-0174-z.
174. Wang X, Kustov S, Van Humbeeck J (2018) A Short Review on the Microstructure, Transformation Behavior and Functional Properties of NiTi Shape Memory Alloys Fabricated by Selective Laser Melting. Materials 11(9):1683. https://doi.org/10.3390/ma11091683.
175. Ullakko K, Huang J, Kantner C, O'Handley R, Kokorin V (1996) Large magnetic-field-induced strains in Ni2MnGa single crystals. Appl Phys Lett 69(13):1966–1968. https://doi.org/10.1063/1.117637.
176. Sozinov A, Lanska N, Soroka A, Zou W (2013) 12% magnetic field-induced strain in Ni-Mn-Ga-based non-modulated martensite. Appl Phys Lett 102(2):021902. https://doi.org/10.1063/1.4775677.
177. Aaltio I, Soroka A, Ge Y, Söderberg O, Hannula S (2010) High-cycle fatigue of 10M Ni–Mn–Ga magnetic shape memory alloy in reversed mechanical loading. Smart Materials and Structures 19(7):075014. https://doi.org/10.1088/0964-1726/19/7/075014.
178. Smith A, Tellinen J, Ullakko K (2014) Rapid actuation and response of Ni–Mn–Ga to magnetic-field-induced stress. Acta Mater 80:373–379. https://doi.org/10.1016/j.actamat.2014.06.054.
179. Gaitzsch U, Pötschke M, Roth S, Rellinghaus B, Schultz L (2009) A 1% magnetostrain in polycrystalline 5M Ni–Mn–Ga. Acta Mater 57(2):365–370. https://doi.org/10.1016/j.actamat.2008.09.017.

180. Chmielus M, Zhang X, Witherspoon C, Dunand D, Müllner P (2009) Giant magnetic-field-induced strains in polycrystalline Ni–Mn–Ga foams. Nat Mater 8(11):863–866. https://doi.org/10.1038/nmat2527.
181. Kok Y, Tan X, Wang P, Nai M, Loh N, Liu E, Tor S (2018) Anisotropy and heterogeneity of microstructure and mechanical properties in metal additive manufacturing: A critical review. Mater Des 139:565–586. https://doi.org/10.1016/j.matdes.2017.11.021.
182. Waske A, Dutta B, Teichert N, Weise B, Shayanfar N, Becker A, Hütten A, Hickel T (2018) Coupling Phenomena in Magnetocaloric Materials. Energy Technol 6(8):1429–1447. https://doi.org/10.1002/ente.201800163.
183. Brown G (1976) Magnetic heat pumping near room temperature. J App Phys 47(8):3673–3680. https://doi.org/10.1063/1.323176
184. Lyubina J (2017) Magnetocaloric materials for energy efficient cooling. J Phys D App Phys 50(5):053002. https://doi.org/10.1088/1361-6463/50/5/053002.
185. Weiss P, Piccard A (1917) Le phénomène magnétocalorique. J Phys Théor App 7(1):103–109. https://doi.org/10.1051/jphystap:019170070010300.
186. Franco V, Blázquez J, Ipus J, Law J, Moreno-Ramírez L, Conde A (2018) Prog Mater Sci 93:112–232. https://doi.org/10.1016/j.pmatsci.2017.10.005.
187. Yang Y, Tong C, Zhong J, Huang R, Tan W, Tan Z (2017) An effective thermal therapy against cancer using an E-jet 3D-printing method to prepare implantable magnetocaloric mats. J Biomed Mater Res B 106(5):1827–1841. https://doi.org/10.1002/jbm.b.33992.
188. Fähler S, Rößler U, Kastner O, Eckert J, Eggeler G, Emmerich H, Entel P, Müller S, Quandt E, Albe K (2011) Caloric Effects in Ferroic Materials: New Concepts for Cooling. Adv Eng Mater 14(1–2):10–19. https://doi.org/10.1002/adem.201100178.
189. Fähler S, Pecharsky V (2018) Caloric effects in ferroic materials. MRS Bull 43(4):264–268. https://doi.org/10.1557/mrs.2018.66.
190. Waske A, Gruner M, Gottschall T, Gutfleisch O (2018) Magnetocaloric materials for refrigeration near room temperature. MRS Bull 43(4):269–273. https://doi.org/10.1557/mrs.2018.69.
191. Lozano J, Capovilla M, Trevizoli P, Engelbrecht K, Bahl C, Barbosa J (2016) Development of a novel rotary magnetic refrigerator. Int J Refrig 68:187–197. https://doi.org/10.1016/j.ijrefrig.2016.04.005.
192. Benedict M, Sherif S, Beers D, Schroeder M (2016) Design and performance of a novel magnetocaloric heat pump. Sci Technol Built En 22(5):520–526. https://doi.org/10.1080/23744731.2016.1185889.
193. Park I, Kim Y, Jeong S (2013) Development of the tandem reciprocating magnetic regenerative refrigerator and numerical simulation for the dead volume effect. International Journal of Refrigeration 36(6):1741–1749. https://doi.org/10.1016/j.ijrefrig.2013.03.012.

194. Engelbrecht K, Bahl C, Nielsen K (2011) Int J Refrig 34(4):1132–1140. https://doi.org/10.1016/j.ijrefrig.2010.11.014.
195. Kolano R, Kolano-Burian A, Hreczka M, Polak M, Szynowski J, Tomaka W (2016) Magnetocaloric Cooling Device with Reciprocating Motion of the Magnetic Field Source. Acta Phys Pol A 129(6):1205–1209. https://doi.org/10.12693/aphyspola.129.1205.
196. Romero GJ, Ferreiro GR, Carbia CJ, Romero GM (2013) Experimental analysis of a reciprocating magnetic refrigeration prototype. Int J Refrig 36(4):1388–1398. https://doi.org/10.1016/j.ijrefrig.2013.01.008.
197. He X, Gong M, Zhang H, Dai W, Shen J, Wu J (2013) Design and performance of a room-temperature hybrid magnetic refrigerator combined with Stirling gas refrigeration effect. Int J Refrig 36(5):1465–1471. https://doi.org/10.1016/j.ijrefrig.2013.03.014.
198. Christiaanse T, Brück E (2014) Proof-of-Concept Static Thermomagnetic Generator Experimental Device. Metall Mater Trans 1(1):36–40. https://doi.org/10.1007/s40553-014-0006-9.
199. Gottschall T, Skokov K, Fries M, Taubel A, Radulov I, Scheibel F, Benke D, Riegg S, Gutfleisch O (2019) Making a Cool Choice: The Materials Library of Magnetic Refrigeration. Adv Energy Mater 9(34):1901322. https://doi.org/10.1002/aenm.201901322.
200. Gutfleisch O, Gottschall T, Fries M, Benke D, Radulov I, Skokov K, Wende H, Gruner M, Acet M, Entel P, Farle M (2016) Mastering hysteresis in magnetocaloric materials. Philos Trans Royal Soc A 374(2074):20150308. https://doi.org/10.1098/rsta.2015.0308.
201. Pulko B, Tušek J, Moore J, Weise B, Skokov K, Mityashkin O, Kitanovski A, Favero C, Fajfar P, Gutfleisch O, Waske A, Poredoš A (2015) Epoxy-bonded La–Fe–Co–Si magnetocaloric plates. J Magn Magn Mater 375:65–73. https://doi.org/10.1016/j.jmmm.2014.08.074.
202. Liu J, Gottschall T, Skokov K, Moore J, Gutfleisch O (2012) Giant magnetocaloric effect driven by structural transitions. Nat Mater 11(7):620–626. https://doi.org/10.1038/nmat3334.
203. Salazar-Jaramillo D, Álvarez-Alonso P, Lázpita P, Sánchez LJ, Gorría P, Blanco J, Chernenko V (2018) Magnetocaloric Effect in Specially Designed Materials. In: El-Gendy AA, Barandiarán JM, Hadimani RL (ed) Magnetic Nanostructured Materials. Elsevier, Amsterdam, p 199–244. https://doi.org/10.1016/B978-0-12-813904-2.00007-3.
204. Moore J, Klemm D, Lindackers D, Grasemann S, Träger R, Eckert J, Löber L, Scudino S, Katter M, Barcza A, Skokov K, Gutfleisch O (2013) Selective laser melting of La(Fe,Co,Si)13 geometries for magnetic refrigeration. J Appl Phys 114(4):043907. https://doi.org/10.1063/1.4816465.
205. Loh G, Pei E, Harrison D, Monzón M (2018) An overview of functionally graded additive manufacturing. Addit Manuf 23:34–44. https://doi.org/10.1016/j.addma.2018.06.023.

206. Bandyopadhyay A, Heer B (2018) Additive manufacturing of multi-material structures. Mater Sci Eng R Rep 129:1–16. https://doi.org/10.1016/j.mser.2018.04.001.
207. Pfaff A, Jäcklein M, Hoschke K, Wickert M (2018) Designed Materials by Additive Manufacturing—Impact of Exposure Strategies and Parameters on Material Characteristics of AlSi10Mg Processed by Laser Beam Melting. Metals 8(7):491. https://doi.org/10.3390/met8070491.
208. Zheng X, Smith W, Jackson J, Moran B, Cui H, Chen D, Ye J, Fang N, Rodriguez N, Weisgraber T, Spadaccini CM (2016) Multiscale metallic metamaterials. Nature Mater 15(10):1100–1106. https://doi.org/10.1038/nmat4694.
209. Mahmoud D, Elbestawi M (2017) Lattice Structures and Functionally Graded Materials Applications in Additive Manufacturing of Orthopedic Implants: A Review. J Manuf Mater Process 1(2):13. https://doi.org/10.3390/jmmp1020013.
210. Tofail S, Koumoulos E, Bandyopadhyay A, Bose S, O'Donoghue L, Charitidis C (2018) Additive manufacturing: scientific and technological challenges, market uptake and opportunities. Mater Today 21(1):22–37. https://doi.org/10.1016/j.mattod.2017.07.001.
211. Dehoff R, Kirka M, Sames W, Bilheux H, Tremsin A, Lowe L, Babu S (2014). Site specific control of crystallographic grain orientation through electron beam additive manufacturing. Mater Sci Tech 31(8):931–938. https://doi.org/10.1179/1743284714Y.0000000734.

Robotics in Manufacturing—The Past and the Present

Ming Li, Andrija Milojević, and Heikki Handroos

1 Robots—From Myth to an Industry Favorite

Where robots are a household name today the idea and concept of robots dates back to the ancient times. Myths mention artificial people, automatons, and mechanical servants programmed to complete tasks and to serve the Gods. According to ancient Greek myth from 400 BC, Hephaestus, the Greek god of invention and technology, constructed the first automaton (humanoid robot) and called it Talos (see Fig. 1)—this gigantic mechanical warrior in bronze was to guard of the island of Crete [3, 4]. The automaton had great speed (250 km/h), the strength to throw boulders, and an ability to heat-up and crush-burn enemies against his red-hot bronze chest. The mythical Talos was powered by the life-fluid "ichor" that ran in a single artery spanning from the neck to a bolt nailed in the warrior's ankle, the removal of which would cause the life-fluid to flush out and Talos to perish—as the myth would have it that would indeed prove to be the demise of this, perhaps the first, robot.

M. Li (✉) • A. Milojević • H. Handroos
Laboratory of Intelligent Machines, LUT-University, Lappeenranta, Finland
e-mail: ming.li@lut.fi; andrija.milojevic@lut.fi; heikki.handroos@lut.fi

M. Collan, K.-E. Michelsen (eds.), *Technical, Economic and Societal Effects of Manufacturing 4.0*,
https://doi.org/10.1007/978-3-030-46103-4_4

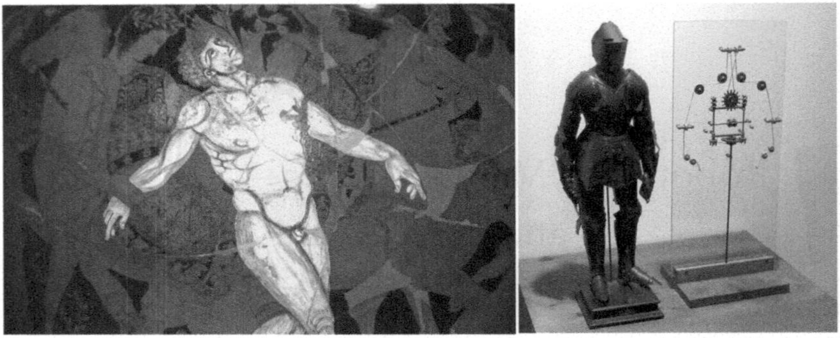

Fig. 1 Detail from an antique vase depicting Talos, (left) [1], Leonardo's automa cavaliere [2]

What the story of Talos and the other ancient myths convey is a story of the ability of humanity to imagine engineered servants or robots already long before our time. One to take the concept further was none other than the great uomo universale Leonardo da Vinci who around the year 1495 designed and constructed a mechanical system called "automa cavaliere" or the mechanical knight. Consisting of cables, pullies, gears, and wheels the knight could be made to perform human-like movement [5]. Based on Leonardo's notes (found only in 1957) a replica of the knight was constructed and the historical accounts on how the original had functioned were corroborated, see Fig. 1.

Until the early twentieth century, mechanical devices with "robotic abilities" were called automatons, or automated mechanical devices—the word "robot" was introduced only in 1920 by the Czech playwright Karel Čapek. In his play "Rossum's Universal Robots" or R.U.R (see Fig. 2), Čapek presents a story of how artificial replicas of humans that have everything except for the soul are constructed to do work that humans don't want to do [7]. Čapek called these artificial men "roboti" based on the Czech word for work. It is said that his original idea was to call them "labori", originally from the latin word "labor" for labor, but that his brother suggested "roboti" and that stuck. At the end of the play the robots rebel against their human masters and finally two robots find love and set off in the sunset to create a new world. Romantic and doomsday-prophetical, the theme has a lot in common with the modern discussion of

Fig. 2 A scene from the play Rossum's Universal Robots. Three robots are shown on the right [6]

ethics of AI. The word robot was universally adopted and is today used to describe autonomous and automatic machines, especially ones that resemble humans in form.

Although the history of robots is interesting and many visionaries have tried to develop humanoid robots, the most impactful applications of robotics have been in industrial manufacturing, where industrial robots and robotic systems are widely used. The first industrial robots were instances of the robotic arm that depending on the design have several degrees of freedom and that can execute predefined tasks. Industrial robots have dramatically changed the workplace, production processes, and manufacturing, as they have replaced humans in performing many tiresome, dangerous, and repetitive tasks—the change is especially pronounced in areas, where human labor was intensive, harsh, or not even possible. The introduction of robots has generally increased productivity, precision, flexibility, and the quality of production.

The first commercially available and programmable industrial robot was developed by George Devol in 1954 and was called "Unimate" [8]. It was a hydraulics-based manipulator arm that could perform different repetitive tasks. The Unimate had four degrees of freedom, 3 rotations, and 1 translation and it could lift up to 34 kg, withstand extreme heat, toxic gases, and radio waves. The robot was sold, e.g., to General Motors and implement on their assembly line in 1961. The robot task was to transport cast parts from an assembly line and weld them onto car bodies, which was a dangerous task for human workers. What followed was a massive adoption of robots into the automotive industry that ignited the establishment of many new robotics companies. In Europe, ABB and KUKA Robotics brought industrial robots to the markets in 1973. In 2017 there were 2.5 million robotic units operational worldwide, where the largest number of industrial robots are active in the automotive industry (126,000 units), followed by the electronics industry (121,000 units), and by the metal (45,000 units), plastic and chemical (20,000 units), and food and beverage (10,000 units) industries) [9]. It is estimated by the Industrial Robotics Federation that by 2021 there will be 3.8 million operational industrial robotic units [10].

A shift in the robotics industry has taken place with introduction of Collaborative Robots or "Cobots". A collaborative robot is an industrial robot that is intended to work alongside humans, in occasions, where there is no danger that the robot will hurt the human. The introduction of cobots is something new and for the first-time robots can work alongside humans and "extend" human capabilities, while not aiming to replace them—as cobots are complementary to humans they will most likely drastically shift and change the future of the workplace.

2 New Trends in Industrial Robotics—Cobots and Advanced Logistics Robots

With the fast evolution of artificial intelligence technology a new types of industrial robots have emerged to supplement and challenge the on-the-grid and relatively inflexible capabilities of standard industrial robots—intelligent cobots supplement and occupy the flexible manufacturing market. The concept of "cobot" was initially formed from General Motors as early as 1994, and it was literally invented in 1996 by J. Edward Colgate and Michael Peshkin and defined as "an apparatus and method for direct physical interaction between a person and a general purpose manipulator controlled by a computer" in a US patent [11].

Today's production lines are agile with emergence of various customized services and must change timely to the fast-varying market. Standard traditional robots are expensive and less versatile in such a scenario and may be out of the reach especially for small and medium-size enterprises that most often require relatively low cost, friendly to use, agile, and fast deployable robots. The intelligent cooperative robots that enter the market can be taught intuitively by operators and can be deployed fast without specific robotic expertise. The range of applications for cobots is wide, from the automotive to the electronics industry, from metal fabrication to packaging and to plastics automation.

In 2008 the UR5 cobot was introduced by Universal Robots in Denmark and in Germany and has rapidly expanded to European and Asian markets. The UR family also has the robots UR3 and UR10 with different payload capacities—all UR robots have an easy programming interfaces and fast set-up, the robots include intuitive and 3D-visualized operations. UR cobots have a high repeatabie ±0.1 mm position accuracy and they have been applied in a wide range of industrial context. In the presence of coworkers the UR robots can adapt to reduced speeds and even make safety stops with the help of sensors.

In 2011, ABB unveiled a twin-arm cobot, FRIDA (Friendly Robot for Industrial Dual-Arm), originally built for the consumer electronics industry and based on customer desire for a robotic solution for manufacturing environments, where robots and humans must cooperate. A later evolved version of FRIDA was renamed as YuMi (You and Me) and officially introduced to the market by ABB in 2015. The YuMi cobot, with flexible hands, parts feeding systems, and camera-based part location ability is suitable for small parts assembly in a small space, collaborating with human workers. The state-of-the-art control algorithms developed for the cobot can pause its motion within milliseconds, when it encounters an unexpected object—or at even a slight contact with a coworker. The position repeatability of the latest YuMi IRB1400 is ± 0.02 mm with payload of 0.5 Kg.

In 2012, "Baxter", the first two-arm collaborative cobot was introduced by rethink robotics, and is usable for a wide range of tasks from packaging and kitting to line loading, machine tending and material handling, and works safely and interactively with coworkers without any caging. A behavior-based user-interface enables the cobot to be programmed intuitively by non-engineers in a matter of minutes. Baxter can also adapt on its own to changes in position and lighting, and to differently shaped objects. Smart sonar systems around the robot head can detect movements within Baxter's proximity, which also enables the robot to learn from its environment. The research version of the robot runs Linux and ROS (robot operation system), which are open source and allow further researcher on many aspects of the robot. The later released single-arm cobot, "Sawyer", retains most of the advantages introduced in Baxter, while presenting improved payload performance and accuracy and the ability to execute tasks impractical to automate with traditional industrial robots.

After these early cobots many robot manufacturers have released a number of cobot designs that are more advanced in terms of their ability to monitor their surroundings (typically more cameras and sensors), in the precision of the tasks that they are able to perform (position repeatability), and in their ability to lift higher payloads. Also the application areas of cobots are increasing and their multi-purpose nature is enhanced with the use of smarter and smarter learning technologies that allow deployment and "task training" to be faster in the real-environment.

Logistics robots have for a long time been considered to be outside the scope of industrial robotics, but they are an important part in the complex and dynamic systems of international trade, of which industry is a part of. Successful applications of logistics robots, such as the autonomous pick and place robots by, for example, Amazon and DHL in their warehouses have greatly increased efficiency in order picking and other warehouse tasks. Even in very large warehouses and their dynamic environments the workflows can be set up and modified quickly with the help of intelligent autonomous logistic robots and robotic data cloud systems. International eCommerce and logistics companies that own and operate warehouses have diversified into robotics by acquiring some logistics robotics companies. It is visible that robotics have become a source of competitive advantage in logistics and the optimization of the use of logistics robots is a way to further enhance the productivity of robotic systems these include, among others, optimization and coordinated autonomy of logistics robots. The management of the robot-fleet is tied with the overall management of a warehouse and tied with demand forecasting based on machine learning algorithms—advanced multi-modal systems allow for real-world order picking to start taking place before a customer has even finished making her selections in a virtual eStore.

The next generation logistics robots are envisaged to be vision guided AGVs (Automatic Guided Vehicles) and "perception" and the ability to grasp objects are perhaps the main areas where development is needed—artificial intelligence combined with ranging sensors and cameras will enable robots to identify, classify, and grasp objects from the best direction and in the best position. Unlike the classic AGV systems that rely on the physical path guidance in the form of embedded magnets, wires, painted lines, magnetic tapes, reflectors, or other path-defining means, the SLAM (simultaneous localization and mapping) algorithms have are used by many modern robots—these robots can autonomously create a map of an unknown environment and maintaining knowledge of their own location within the created map. Typically the mapping is achieved through scanning with a 2D or a 3D Lidar, often supported with a 3D stereo depth camera with advanced sensing—combining the odometer and some advanced filtering techniques allows the robots to estimate quite precisely their position on the map, while being able to avoid unknown (even moving) obstacles on their path. Logistics robots with this kind of capability are able to operate autonomously also in locations "they have never seen before"—unsurprisingly similar technologies are commonly used in robotic vacuum cleaners in household use. It can be seen that there is a merger of robotics with artificial intelligence going on that will result in more precisely and autonomously functioning robots that are multi-usable. In fact, sharing and copying the characteristics that have been found to be successful in the context of logistics robots in the large e-shopping warehouses, large manufacturing factories are also adopting logistics robots to support manufacturing operations.

Where many cobots are fully autonomous, also remotely operated systems can be considered cobot systems. There are still many places where humans are needed to operate machines and systems that require more cognitive skills than the automation of today can provide. In the future, the emerging wireless communication technologies will be able to provide a sufficiently small latency to carry out work tasks that require multisensory feedback to the remote operator—making such fast data transfer available globally may significantly change the labor markets at least in some niche areas, as remote operation of devices and machines by skilled operators can be done from anywhere in the world. Fast 3D camera-technology, virtual helmets, and haptic interfaces may someday provide the remote operator with a close to a fully realistic feeling of presence in the actual machine performing the task.

All in all it can be observed that industrial robotics is a highly interdisciplinary area that covers many fields from mechanical-, electronics-, and dynamics design, to construction of actuators and servo driving technology, to signal processing and control, and to AI and software development.

3 Supporting Technologies Behind Modern Robotics

Machine vision has now become an important integral part of perception ability of an intelligent industrial robot, which enables a robot to recognize objects it is handling and to position the end-effector through in-hand vision, and to perceive and to construct a cooperative environment for human coworkers. *Convolutional neural networks* play an important role in how machine vision is put into practice today—they are used already quite commonly in image classification and in object detection and tracking. These skills are interesting, for example, when industrial robots are used in product quality control.

As one of the targets for industrial robots is to function in a "partner" role with human coworkers, while being autonomous, *reinforcement learning* is an important tool in enabling robots to learn "behavior" through interaction [12]. A variety of complex problems, where no obvious programmable pattern and behavior appears to be discernible due to the environmental dynamics or uncertainty, can be solved with reinforcement learning—such problems include force control [10] used, for example, in robotic hands and as technology to optimize robots grasping things and for motion planning [13–15].

Historically artificial neural networks have been applied in robotics for a long time, mostly via their applications to various control problems. One direction, where the study of artificial neural networks has gone in the context of robotics is *reservoir computing* [15].

Evolutionary algorithms have become an important search and optimization technique within the last two decades, since robotics contains high dimensional parameters that lead to identification and optimization problems, evolutionary algorithms have been actively applied to robotics, covering areas such as kinematic and dynamic parameter identification, controller parameter optimization, trajectory generation, and motion planning. Evolutionary algorithms are just one type of optimization tool however, for one reason or another, most likely the malleability of these methods fit many types of problems, and they have been often used in the context of robotics. Typically the optimization is done off-line due to the computational cost, which means that evolution of a robot in terms of optimization and re-optimization is step-wise—in the future real-time re-optimization will become possible and the advanced sensors that provide robots with high quality real-time situational awareness can be used to generate optimization problems that are solved instantly and the robot acts accordingly.

4 Conclusions

Although the origins of the industrial robot were in the field of automobile assembly, its application has expanded into virtually all fields of manufacturing and process industries. With the advances made in artificial intelligence that translate to better ability to control the physical functionality of robots and sensor technology that allow robots to acquire sensory skills that match those of humans it can be forecasted that intelligent robots may become ubiquitous is manufacturing, and elsewhere in the society. Robots that have learning capabilities may not only boost the productivity over a wide variety of application areas in manufacturing, but may also provide a basis for using robots in tasks where they have not previously been used. A learning robot that learns in a place where it originally was not designed to learn can be said to be a step closer to an all-purpose robot and becomes a "method of invention". Importantly, modern industrial robots are not any more limited by their ability working on a single task alone, but possess abilities that allow them to be used in connection with various different task the level of difficulty of which is increasing. The enhanced learning abilities present in modern intelligent robots are an important step in terms of general applicability of robots—looking at this situation from the point of view of economics, a robot with the capability to handle a large number of tasks that can easily be taught to handle "any compatible task" has the advantage and potential of becoming a standard tool for various tasks. In this vein, it is not wrong to expect that manufacturing will continue to be disrupted by robotics in the future. On the other hand, it should be remembered that technological developments in robotics and automated manufacturing systems don't take place in a vacuum. It is too simple to argue that advancements in technology shape social structures and social potential. The critical history of robotic and automated manufacturing systems demonstrates an opposite reality. Technological potential embedded in industrial robots and manufacturing systems has often been limited by social constrains. Efforts to replace human labor and to create industrial processes that exist without people, have been hampered by political, economic, and ideological tensions. It has been important to soften the disruption and to present that robots and automated manufacturing systems are "collaborators". Machines will take over human work and perform difficult and dangerous tasks, but the ultimate command will be in the hands of skilled workers and managers. It is difficult or perhaps even impossible to predict, for how long human labor is needed. Technology optimists suggest that the power has already shifted and the advanced systems are taking over also managerial tasks. The opposite argument emphasizes typically historical experience. What can be said is that social structures within manufacturing industries are strong and resilient, while the human mind is able to cope with even rapid technological advances.

References

1. Photo: Forzaruvo94 / CC BY-SA (https://creativecommons.org/licenses/by-sa/3.0).
2. Photo by Erik Möller. Leonardo da Vinci. Mensch—Erfinder—Genie exhibit, Berlin 2005./Public domain.
3. A. Mayor, "Gods and Robots: Myths, Machines, and Ancient Dreams of Technology", Princeton University Press, 2018.
4. R. Apollonius, "The Argonautica", CreateSpace Independent Publishing Platform, 2017.
5. R. Mark, "Leonardo's Lost Robots", Springer Berlin Heidelberg, 2016.
6. Photo: Unknown, uploaded to ar.wiki by ar: Classic 971 on 13 July 2006, to en.wiki by en:User:Rocketmagnet on 4 December 2007, to commons by User:Sreejithk2000 on 2 June 2010/Public domain.
7. K. Capek, "R.U.R.: (Rossum's Universal Robots)", Players Press, 2002.
8. A. Gasparetto, and L. Scalera, From the Unimate to the Delta Robot: The Early Decades of Industrial Robotics" in: Proceedings of the 2018 HMM IFToMM Symposium on History of Machines and Mechanisms. https://doi.org/10.1007/978-3-030-03538-9_23, 2019.
9. Executive Summary World Robotics 2019 Industrial Robots, International Federation of Robotics (IFR), 2019.
10. R. Meyes et al., "Motion Planning for Industrial Robots using Reinforcement Learning," Procedia CIRP, vol. 63, pp. 107–112, 2017.
11. "Cobots" US Patent 5,952,796, https://patents.google.com/patent/US5952796.
12. F. Stulp, E. Theodorou, J. Buchli, and S. Schaal, "Learning to grasp under uncertainty," Proc.—IEEE Int. Conf. Robot. Autom., pp. 5703–5708, 2011.
13. M. Vecerik, O. Sushkov, D. Barker, T. Rothorl, T. Hester, and J. Scholz, "A practical approach to insertion with variable socket position using deep reinforcement learning," Proc.—IEEE Int. Conf. Robot. Autom., vol. 2019-May, pp. 754–760, 2019.
14. Y. Litvak et al. "Learning a High-Precision Robotic Assembly Task Using Pose Estimation from Simulated Depth Images." ArXiv abs/1809.10699 (2018): n. pag.
15. A.S. Polydoros, L. Nalpantidis and V. Kruger, "Advantages and Limitations of Reservoir Computing on Model Learning for Robot Control", Paper presented at IROS Workshop on Machine Learning in Planning and Control of Robot Motion, Hamburg, Germany.

Maintenance-Management in Light of Manufacturing 4.0

Michele Urbani, Dario Petri, Matteo Brunelli, and Mikael Collan

1 Introduction

During the last decade, the manufacturing industry has gone through a deep transformation with the digitalization of processes, the arrival of the Internet of Things, the spread of artificial intelligence (AI) in daily practices, and the ubiquitous presence of data—thanks to the cloud technologies lifting the efficiency of manufacturing systems to a new level. Notwithstanding these radical changes, the manufacturing industry still has a strong dependence on maintenance, a field that is still considered to be a *necessary evil* by most managers, but without which plants and equipment will not remain safe and reliable. The importance of

M. Urbani (✉) • D. Petri • M. Brunelli
Department of Industrial Engineering, University of Trento, Trento, Italy
e-mail: michele.urbani@unitn.it; dario.pertri@unitn.it; matteo.brunelli@unitn.it

M. Collan
School of Business and Management, LUT University, Lappeenranta, Finland
e-mail: mikael.collan@lut.fi

M. Collan, K.-E. Michelsen (eds.), *Technical, Economic and Societal Effects of Manufacturing 4.0*,
https://doi.org/10.1007/978-3-030-46103-4_5

maintenance-management as part of tangible asset-management is clearly inscribed within modern international industry standards [1], where asset-management is defined as "the coordinated activity of an organization to realize value from assets". Maintenance-management takes care of physical assets with the aim of minimizing their life-cycle cost and achieving stated business objectives. Depending on the specific sector of industry, maintenance takes different forms—its most elementary form involves simple operations and inspections of and on machines, while the most cutting-edge applications include intelligent maintenance control-systems capable of predicting the remaining useful life (RUL) of components and triggering maintenance activities automatically when needed. Moreover, some companies are adopting more holistic approaches to maintenance, aimed at improving the efficiency of the whole productive unit. Such approaches are called *total productive maintenance* (TPM) [2] and they aim at improving the quality of products, developing corporate culture, and enhancing the attention to safety and environment.

The popularization of Industry 4.0 paradigm around the year 2011 represented a new starting point for the manufacturing industry after the financial crisis of 2008. Asset-management and maintenance-management of physical equipment underwent a transformation: real-time monitoring of working conditions became very common due to decreasing cost of sensor technology (IoT devices), thus making possible the development of new technologies such as Virtual Factories and Digital Twins (DTs) of machines and processes. The digital replication of the physical environment allows the optimization of processes already during the design phase and the optimization of running processes during the production phase. Real-time monitoring of assets and the direct control of processes remotely has became a part of the new paradigm of manufacturing; with respect to maintenance, diagnostics and prognostics of equipment are spreading into daily practices and a new stream of research is contributing to the development of these technologies.

In this chapter we illustrate some of the connections between modern manufacturing (Manufacturing 4.0) and maintenance-management, present shortly the evolution of maintenance-methodologies starting from early models until today and summarizing the most important concepts relevant to the field including a discussion of how the digital twin concept may become an important issue for maintenance-management.

2 Maintenance-Management: An Overview

Maintenance-management is nowadays a fundamental function in most industry. In its traditional form, maintenance is aimed at ensuring that a system performs its function in a safe and efficient manner. Due to information technology (IT) development, maintenance-management has seen a significant evolution within its best practices: the classical methods for maintenance-planning and scheduling have been integrated and improved by technologies such as the Internet of Things, cloud computing, and artificial intelligence.

Engineering systems often have a complex structure, with a limited number of dedicated resources and strict requirements on safety and on performance—under these circumstances maintenance is an issue that needs to be handled in a systematic way. A clear strategy for maintenance must be defined, where components of a system to be maintained should be documented and listed according to priority, then a set of rules for the daily management of operations must be drafted. The set of rules that are used to coordinate maintenance tasks are typically called a *maintenance-policy*. As basic example, maintenance-policies for lifts and elevators that typically depend on country-wise regulations and that state that maintenance must be carried out on regular intervals, such as "every twelve months", which is then the rule that triggers a maintenance intervention that is aimed at avoiding sudden failures of the system. The above discussed types of interventions that are carried out before a failure has taken place are called *preventive* and they may range from simple inspections to the replacement of broken components. Maintenance actions undertaken after a failure are called *corrective* and they typically consist of the replacement and/or the repair of failed components. Usually corrective actions are more expensive than preventive, but when this is not the case it is sometimes possible to let a system *run to failure* that is, a system is left un-serviced until it fails, or until its fails and its failure is detected. Non-critical system components with a steady failure rate are often let run to failure.

Implementing preventive maintenance-policy typically requires more in terms of analysis, than a corrective policy—it requires information about the state of the maintained system such as information about the degradation level of system components. Depending on the information available, preventive maintenance-policies can be time-, or condition-based.

Time-Based Maintenance

Time-based or predetermined, as they are also called, maintenance-policies were the first approach adopted to effectively manage maintenance. In these types of policies maintenance actions are scheduled to take place on predefined times, according to set intervals of duration t_M, or upon failure (whichever occurs first). The aim of the policies is to preventively maintain the asset through shorter, but planned downtimes and by doing so avoiding longer and more expensive corrective maintenance actions. In this way the asset availability increases and consequences of failure can most often be avoided.

Scheduling of activities can be organized according to block-based- or age-based approaches. Block-based approaches schedule maintenance actions at constant time intervals, regardless of the asset operating time. The block-based approach is commonly used, when several assets of the same class (a block) are in (constant) use simultaneously. Age-based, or runtime, models are applied, when asset degradation and failures depend on the cumulative load exposure. Since the active age of a mechanical component has a strong correlation with the physical wear, or fatigue, of a component the maintenance of mechanical systems is often managed according to the age of system components. Asset age can be measured by using the working time of a machine as proxy, or in other ways, such as by observing the number of kilometres travelled or by the number of take-offs or landings, as can be done with aircraft. Approaches that combine more than one proxy for component states are also possible. Literature is ripe with research on time-based approaches for maintenance-optimization, we refer the interested reader to see the review by Wang [3]. It is worth to mention that time-based maintenance-policies carry a risk of over-maintenance, as some of the performed actions may not be necessary, on the other hand, time-based policies cannot weed-out failures, when component-deterioration happens at a non-standard pace—these are clear handicaps, when compared to condition-based policies. In fact, when the cost-risks of a time-based policy, or the costs of over-maintenance, are too high, condition-based maintenance may represent a feasible alternative.

Condition-Based Maintenance

Experience shows that failures can occur independently of the asset age, but at the same time most of these undesired events give some sort of warning about the fact that they are about to occurring—thanks to the presence of such symptoms an early detection of fault occurrence is possible. This means that preventive actions can be taken, if the signals and symptoms of impending failures are understood, this is the fundamental concept that underpins condition-based maintenance. According to condition-based maintenance-policies maintenance actions are initiated by performance of a system reaching a trigger-level, typically determined by monitoring one or more indicators (sensors) of the maintained system. This means that maintenance is not done based on a predetermined schedule, but actions are taken based on observed, evidence-based deterioration of system performance that signals impending (component) failure and as such on only-when-needed basis.

A prerequisite for condition-based maintenance-policies (CBM) is that the there is objective ***monitoring*** of the system state in place—the monitoring should be carried out in a non-invasive way and it is typically achieved by using sensors. Monitoring can be scheduled or continuous and the output from monitoring is a set of observations (indicators, failure precursors) that describe the capacity of a system to perform its function. A typical example of a failure precursor is the vibration frequency of a rotating machine—shift in the frequency is a clear indication of a change in the working conditions. As a rule of thumb used in CBM, once enough data has been gathered, thresholds on the monitored feature-values are established to more reliably identify degraded asset performance—a comparison between the system-state and the thresholds is used to track the system health. With knowledge about the system health and history-based thresholds a decision about maintenance-scheduling can be made in a way that actions are performed only when needed and as a result both the probability of failure and the overall cost of maintenance can be optimized.

3 More About Condition-Based Maintenance

Setting up condition-based maintenance is a process and it can be divided roughly into three main steps. Condition-based maintenance assumes that objective monitoring of the system is possible, which means that ***acquisition of data*** about the system state is in place. Sensors that measure issues such as material cracking, corrosion, vibration, and change in electrical resistance are the types of information that are usable from the point of view of understanding the system state—one must also remember that these issues depend on the operating and the environmental conditions, such as the frequency of use, ambient temperature, and humidity. It is typical that a monitored system must be equipped with sensors, signal conditioning and digitizing components that are typically already embedded in new modern machines. We emphasize the importance of sensors, because they are a core technology needed for the implementation of the Manufacturing 4.0 paradigm in maintenance—they are the bond that connects machines into networks and they allow the realization of the Internet of Things.

Based on the data collected the features that explain and describe the state of the system and allow determining whether maintenance is necessary must be estimated. Features can be difficult to observe directly (by observing the system), but by exploiting data and a priori knowledge of the system *feature extraction* can be made easier. The quality of a feature is determined by its capacity to represent the system state, in order to achieve a better state representation, usually a set of features is used—the more clearly different system states can be distinguished from each other the better the condition of the system can be described. In practice finding the correct features or sets of features that allow high failure detection capability and a low false alarm probability are problems that can be solved by specific methods created for feature-selection and for information fusion. Improvement in feature-selection methods has been fuelled by the great interest analytics and AI have received in recent times. One must remember that sudden changes in the operative and environmental conditions may render features that work well under normal conditions imprecise—this is why the best modern systems may use different sets of features for different operating conditions and are able to change the feature sets used "on the fly", when conditions change.

Once the data acquisition and feature extraction processes are ready *condition monitoring* can be effectively performed. Monitoring is the last step prior to the definition of the maintenance-strategy that is forming the set of rules that aids managers in taking maintenance decisions.

The main goal of condition monitoring is to provide fault-recognition, which typically foresees three sub-goals: (1) fault detection, aimed at identifying if a fault or the degradation of a component occurred; (2) fault isolation, that identifies the damaged component among many others; and (3) fault identification, aimed at determining the nature, extent, and severity of the isolated fault. In the following we look at these issues in more detail.

Fault Detection

The task of fault detection is to identify the presence of abnormal working conditions in a system by leveraging the information from the system history and information that can be learned from actual data. Typically a benchmark that defines the "normal" working conditions of the system is needed—the normal conditions depend on the task that the system is carrying out and on the environment surrounding the system. Because of different environments a system may have several normals—each normal will have a "profile" that is a set of features that defines it. Another thing is the extraction of profiles for different fault-states, such as "healthy", "degraded", and "faulty". The state of the system can be compared to the different profiles and this allows one to understand the state of the system and to predict the failure. Typically one will want to see several system states that precede the "failed" state, because the more states there are the finer is the information about the system state and better one can predict what will happen next. The comparison of the observed system state and the normal state can be done by different means, two examples of usable modelling techniques for this purpose are the auto associative kernel regression (AAKR) [4] and principal component analysis (PCA) [5] for the identification of the state and subsequently a statistical test is applied to identify the extent to which the state of the system differs from a normal condition. Typically used tests include the threshold based approach, Q statistics, and the Sequential Probability Ratio Test (SPRT) [6]. When the state of the system is known an action is taken (not taken) depending on the recommendations described for each state—the recommendations are drafted by using fault diagnosis techniques.

In order to clarify how fault detection works, we provide a simple example of condition monitoring. We assume that the state of a system is represented by a single feature $x(t)$. We define two thresholds considered important for the component. In Figure 1, the first threshold x_W identifies a warning-level, while threshold x_F identifies the failure of the component. When the value of $x(t)$ surpasses level x_W, an alarm is triggered, and a preventive action can be undertaken to prolong the life of the component, or to change it, to avoid incurring a sudden failure. The curve representing the behaviour of $x(t)$ is known as the Performance/Failure curve and it expresses the evolution of the system-feature as a function of either calendar time or system age time. A realistic mathematical model of $x(t)$ will also include the uncertainty related to the estimated quantity, which in

Fig. 1 A performance/failure curve for a generic system

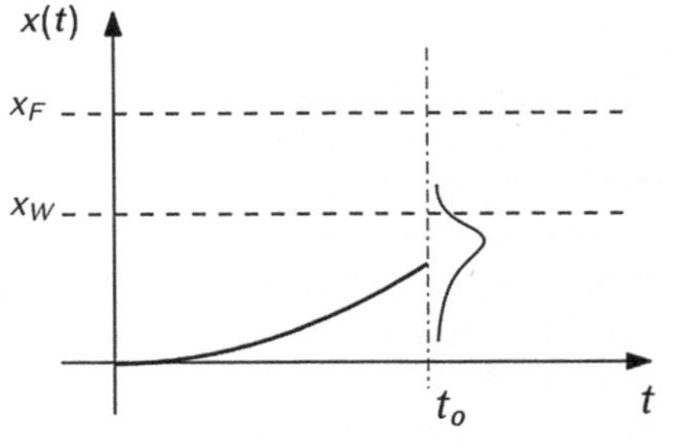

Fig. 1 is represented by a generic probability distribution. The importance of modelling the degradation of a component using a random variable is represented by the possibility to express the result using a probability that is, a degree of belief about the triggering event.

Fault Diagnosis

Fault diagnosis is isolating and identifying the fault and typically means identifying the cause, this means identifying which component in a system is degrading among many possible components and to determine the nature, the extent, and the severity of the fault. Isolating and identifying the fault are sometimes overlapping and not always clearly separable. Fault diagnostics means most often solving a classification problem—any given set of measurements from the system can be matched to a single component if sufficient data is available for training a machine learning classification algorithm. In cases where data is abundant algorithms can even spot specific conditions within components and provide a credible probability of a failure event. Many techniques are good for this task, the interested reader may find an extensive review about modern fault diagnostics techniques applied to rotatory machines in [7], where the authors describe both the fundamental principles behind adopted AI algorithms and present numerous application examples. As a caveat about AI-based techniques one must observe that where there is no data, or data is very incomplete, machine learning algorithms cannot be used—in such cases suitable data must first be collected. In the cases of very rare faults diagnosis is difficult and diagnostics performance for them is typically poor.

The performance of condition-based maintenance systems is only as good as the system in place and there is uncertainty associated with the outputs (alarms) from these systems. Uncertainty is caused by a number of things, some were already mentioned above such as the operating conditions and the environment, but others like production tolerances also affect the reliability of CBM system—because of tolerances two nominally identical machines may have a different wear. Due to this inherent inaccuracy the output from CBM systems is most often expressed as a probability or an interval. We refer the reader interested in deepening their knowledge in maintenance and maintenance optimization to read the review by De Jonge and Scarf [8].

4 Prognostics and Health Management—Towards Industry 4.0

Thanks to the availability of cheap networked sensors the monitoring and maintenance of systems is undergoing a fast and deep change. In the past, manual collection of maintenance-relevant data made the processing slow and unreliable—today technology allows abundant collection of data often in real-time. This profound change has caused the attention of maintenance systems development to move towards maintenance process-optimization. The new generation of production systems that are "smart" and networked has been labelled as Cyber Physical Production Systems (CPPS)—important to maintenance, they offer the possibility to perform real-time monitoring and accurate analysis of the degradation of critical components. This means that the long stream of research carried out on condition-based maintenance can now be exploited for its full potential—this change has given rise to the term Prognostics and Health Management (PHM), which can be said to be the cutting-edge approaches to predictive maintenance born within the last two decades. Keeping in mind that PHM is part of the same continuum with CBM and that the two cannot be sharply separated, it can be said that PHM aims higher than the "traditional CBM" and uses more advanced tools to get there.

The higher goals of PHM include, for example, optimization of maintenance-planning, reduction of downtimes, just in time spare parts provision, energy consumption optimization, minimization of raw material use and of pollution—all in all the focus is on increasing profitability through "better maintenance". PHM means effectively the same thing that is meant when the term Predictive Maintenance is used in common parlance. A fundamental prerequisite for a well-functioning predictive maintenance system is the high quality of information that is used as an input into the system. This is true for both the real-time operation of the system as it is true for the information that is needed to construct or teach the system to be able to operate reliably—the information needed typically includes operating and maintenance histories, prior knowledge about system failure modes, resource constraints, and mission requirements. The information is used in tuning complex models the architecture of which may include numerous machine learning sub-systems and that require top of the line know-how. This means that these systems are expensive and they can be constructed only for systems that either merit such costs from the point of view of safety or that are business-critical and can economically justify the expenses.

In prognostics and health-management systems the system status received as input from condition monitoring is used to create an estimate of the system degradation state, which is used together with the P/F curve, or by using a classification-based architecture, to determine the distance between the current degradation level and a failure threshold (health-margin). The idea of the modern systems is to not only identify the cause of the fault but also to predict any secondary failures that may occur and to forecast the system health evolution as reliably as possible. Prognostics is considered the "holy grail" of PHM systems [9], because diagnostics has a retrospective approach to failure that consists of identifying and quantifying failures that have already occurred, while prognostics is about forecasting and as such, if successful means that the remaining useful life (RUL) of components can be accurately predicted. This will happen simply by being able to accurately estimate the end of life of a component and calculating the time to the end of life—the more accurate this ability is, the more precise can any optimizations performed based on it, including just in time deliveries of spare parts and maintenance scheduling become. The difference between high accuracy and medium accuracy can mean great savings in cases, where multiple systems are maintained and costs associated with maintenance are high. Another important issue is to know how much in advance a prognostics system can (accurately) predict the failure time—in fact, the relative RUL estimation accuracy and the prognostic horizon are key performance parameters of PHM systems.

In the literature, three types of approaches to prognostics have been identified, namely (1) experience-based approaches, which exploit historical information of a similar components; (2) model-based approaches, which make use of a physical fault model, and; (3) data-driven approaches, which are mainly based on AI-techniques. We propose the interested reader to explore model-based and data-driven approaches by reading the book by Kim et al. [10].

Digital Twins and Their Connection to Maintenance

According to recent literature on maintenance and industrial management [11, 12] prognostics and health management systems be viewed as an examples of cyber-physical systems (CPS). The idea of CPS started to spread in the beginning of the 2010's, when NASA published their Modelling, Simulation, Information Technology & Processing Roadmap [13]—the document delineated the intention to integrating all the available physical and virtual technologies, the context back then was aeronautics. In essence the idea is that of a digital replica of a physical asset and it was called a Digital Twin (DT) and defined as "*an integrated multi-physics, multi-scale, probabilistic simulation of a vehicle or system that uses the best available physical models, sensor updates, fleet history, etc. to mirror the life of its flying (sic) twin. It is ultra-realistic and may consider one or more important and independent vehicle systems*". What makes this interesting from the point of maintenance is that predictive maintenance was one of the first fields of application of the DT concept, together with the check of mission requirements and a more transparent life-cycle view. The DT concept was subsequently extended to the manufacturing industry and the term Cyber-Physical Production System (CPPS) was coined to indicate the specific application area. A CPPS is composed of a physical part, a virtual part (the DT), and a stream of data between the two [14]. The DT strives to hold a perfect real-time synchronization between the physical and the virtual worlds, the physical part sends data to the virtual model, and the virtual part reproduces the physical system with ultra-high fidelity. As this is the case, historical data stored can be used together with real-time sensory information from the physical system in order to run, e.g., simulations and to optimize the production process virtually and then transmit "orders" to the physical system in order to optimize the way it functions. Theoretically the CPPS can harness the interaction between the virtual and the physical parts in order to create a continuously improving system. Digital twins are a clear way to remedy the typical problems of data collection, organization, and exploitation widespread in the context of production systems.

In fact, digital twins start to look like the key to reaching solutions for the problems of fitting together the best practices in engineering design and in process control. The advantages of adopting the DT concept seem cover the whole of product lifecycle that is, production design, manufacturing, and service providing are all immersed in the realm of DT [14]. In the design phase, if realized with a sophisticated digital model, issues that have to do with the maintainability of the production system can perhaps be addressed already on the drawing board—this may include the instrumentation of the system for best possible diagnostics and prognostics. During the production life of the production system the DT can perhaps assist in production planning, resource management, and procurement that can be optimized also with regards to predicted downtimes due to maintenance. The DT may run failure prediction algorithms in real-time so that users can be notified when the system state changes and in cases of imminent failure. It seems feasible to say that there is clear potential for maintenance systems development based on the digital twin concept.

5 Conclusion

Maintenance has always been a part of the management of production systems and it has become a craft of its own, the early mathematical models for maintenance management were based on the notion of optimizing the interval between maintenance activities in order to minimize downtime and the maintenance related costs. This type of maintenance management systems may still exist in cases, where preventive maintenance is the norm and the systems maintained are "old school" and not instrumented with sensors.

The modern approach of maintenance management is based on condition-based maintenance, which in the early days was more expensive than time-based maintenance management and thus reserved to high-risk and high-cost applications. Today the price of sensors and instrumentation is considerably low, which has made condition-based maintenance the leading way of handling maintenance management. Improvement of maintenance policies has created competitive advantages for companies that have been able to adopt them successfully and therefore a shift to modern maintenance management approaches is occurring in many companies. Automation of industrial facilities, such as the increasing use of robotics, improves productivity and safety, but it also increases the technological complexity of industrial assets and means a higher dependence on production systems—this accentuates the role of effective and efficient maintenance.

Key Industry 4.0 technologies, such as artificial intelligence and Internet-of-Things, enable the implementation of very effective maintenance policies at an affordable cost and have paved the way for better diagnostic and prognostic systems, which can be said to be the backbone of what is typically called predictive maintenance. These systems are able to make fault-prediction even more accurate than what is possible with traditional condition-based maintenance methods and therefore offer a possibility for even further savings through better optimization. Predictive maintenance most importantly is a forward looking approach to maintenance, where traditionally the policies have been based on after-the-fact optimization.

The concept of digital twin is interesting from the point of view of maintenance management, as it is based on the idea of having a highly accurate real-time virtual model of a physical system that are "conversing" with one another. In effect, this is a concept that is not very far away from the ideal maintenance management system in terms of the information exchange between a production system and the maintenance management system. The digital twin, as it is used in the lifecycle management of products today is already opening avenues for many issues that are relevant to making maintenance better—looking forward there is potential for much more, specifically in terms of using digital twins in a maintenance focused way.

Getting back to the real-world, one must observe that the choice of maintenance management systems and policies is always constrained by the economic and technical realities surrounding the maintained systems. In this respect, predictive maintenance is at the start of a road that may lead at some point to something that resembles a digital twin—one thing is for sure, the Industry 4.0 paradigm and what we already can see beyond it will change maintenance management.

References

1. ISO 55000, "Asset Management—Overview, Principles, and Terminology." 2014.
2. S. Nakajima, *Introduction to TPM: Total Productive Maintenance*. Productivity Press, Inc., 1988.

3. H. Wang, "A survey of maintenance policies of deteriorating systems," *Eur. J. Oper. Res.*, vol. 139, no. 3, pp. 469–489, 2002.
4. P. Baraldi, F. Di Maio, P. Turati, and E. Zio, "Robust signal reconstruction for condition monitoring of industrial components via a modified Auto Associative Kernel Regression method," *Mech. Syst. Signal Process.*, vol. 60, pp. 29–44, 2015.
5. P. Baraldi, G. Gola, E. Zio, D. Roverso, and M. Hoffmann, "Expert Systems with Applications A randomized model ensemble approach for reconstructing signals from faulty sensors," *Expert Syst. Appl.*, vol. 38, no. 8, pp. 9211–9224, 2011.
6. F. Di Maio, P. Baraldi, E. Zio, and R. Seraoui, "Fault detection in nuclear power plants components by a combination of statistical methods," *IEEE Trans. Reliab.*, vol. 62, no. 4, pp. 833–845, 2013.
7. R. Liu, B. Yang, E. Zio, and X. Chen, "Artificial intelligence for fault diagnosis of rotating machinery: A review," *Mech. Syst. Signal Process.*, vol. 108, pp. 33–47, 2018.
8. B. de Jonge and P. A. Scarf, "A review on maintenance optimization," *Eur. J. Oper. Res.*, 2019.
9. J. B. Coble and J. W. Hines, "Prognostic algorithm categorization with PHM challenge application," *2008 Int. Conf. Progn. Heal. Manag. PHM 2008*, 2008.
10. N.-H. Kim, D. An, and J.-H. Choi, *Prognostics and Health Management of Engineering Systems*. Springer, 2017.
11. E. Negri, L. Fumagalli, and M. Macchi, "A Review of the Roles of Digital Twin in CPS-based Production Systems," *Procedia Manuf.*, vol. 11, no. June, pp. 939–948, 2017.
12. W. Kritzinger, M. Karner, G. Traar, J. Henjes, and W. Sihn, "Digital Twin in manufacturing: A categorical literature review and classification," *IFAC-PapersOnLine*, vol. 51, no. 11, pp. 1016–1022, 2018.
13. NASA, "Modeling, Simulation, information Technology & Processing Roadmap—Technology Area 11," *Natl. Aeronaut. Sp. Adm.*, p. 27, 2010.
14. F. Tao, J. Cheng, Q. Qi, M. Zhang, H. Zhang, and F. Sui, "Digital twin-driven product design, manufacturing and service with big data," *Int. J. Adv. Manuf. Technol.*, vol. 94, no. 9–12, pp. 3563–3576, 2018.

PART II

Manufacturing 4.0 Business Models and the Economic Feasibility of Additive Manufacturing

Industrial Additive Manufacturing Business Models—What Do We Know from the Literature?

Jyrki Savolainen and Mikael Collan

1 Introduction

Additive Manufacturing (AM or 3D-printing) is commonly understood as the ability to create parts or products directly from digital blueprints by adding material layer-after-layer [1–9]). AM is one of the key-technologies in the Manufacturing 4.0 paradigm that revolves around cyber-physical systems and small-scale data-driven production [10–13]. In this chapter we focus on business models associated with additive manufacturing that we define as *the logic of creating and capturing value through a series of interdependent activities of which one is additive manufacturing.*

The literature on the business models based on additive manufacturing can be divided roughly into four quadrants, based on the speed and strength of the change imposed by additive manufacturing to the world of manufacturing (disruptive or incremental) and, on the other hand, based

J. Savolainen (✉) • M. Collan
LUT University, Lappeenranta, Finland
e-mail: jyrki.savolainen@lut.fi; mikael.collan@lut.fi

M. Collan, K.-E. Michelsen (eds.), *Technical, Economic and Societal Effects of Manufacturing 4.0*,
https://doi.org/10.1007/978-3-030-46103-4_6

on the openness of the business model adopted by agents (closed or open) [14]. The openness typically is a divisive issue between the open hobbyist additive manufacturing sector and the closed for-profit industrial manufacturing sector. In this chapter we concentrate only on the closed, for-profit industrial sector business models. We omit having a deeper look in the literature that concentrates on small-scale "prosumer" activities, e.g., local print-shops operated by 3D-printing enthusiasts [15], and other user-entrepreneur -based business models, such as AM production design or consulting to other hobbyists [16].

From a production technology point of view the key promise of additive manufacturing lies in the ability of AM to turn physical materials into a desired form without many of the costs one incurs by using conventional manufacturing methods. There is potential for cost savings emanating from AM when machining, molding, casting, and tooling are not required, or there is a remarkable difference to current practice in how much of these costly actions are required [4, 17–22]. One of the drivers of AM is the potential to manufacture certain components or products at a lower cost—this potential has not yet been universally realized, far from it, as AM technology is still in many aspects developmental.

Other technical drivers for the adoption of additive manufacturing may include issues such as the high customization capabilities offered, product quality improvements, e.g., weight reduction and better product geometry, and production flexibility [23]. One obvious, but often unmentioned issue that may drive additive manufacturing technology is its high level of automation. Generally one can observe that the first instances, where additive manufacturing seems to be the most cost efficient way to produce industry-grade components have been found in biomedical, automotive and aerospace industries (see, e.g. [4, 9]).

The diffusion of additive manufacturing technologies is hampered by the limitations to the materials, i.e., filaments that can be used in the process. Generally, AM technology is not far enough to be able to take use of many of the materials available for conventional manufacturing methods, while another material-related issue is the filament unit cost in AM, which is typically higher than the equivalent raw material cost in conventional manufacturing [24]. Also the production speed (rate) of additive manufacturing can be a limiting factor, especially when large components/ structures are additively manufactured [25]. It is also noticeable that contrary to conventional manufacturing, the economies of scale in AM are most often limited, as the cost of raw materials typically have a direct relationship with the production volume [24]. At the time of writing, the majority of additively manufactured products still require surface finishing [2, 4, 22], which makes many of additive manufacturing processes only capable of producing semi-finished products. As a solution for mass production applications several authors, e.g. [3, 24], suggest production layouts combining additive and conventional manufacturing techniques—this means the printing of "near-net-shapes" and machine finishing products with subtractive methods. As with all new manufacturing technologies concerns have been voiced also about the mechanical properties of products manufactured by using AM-technologies [1, 3, 26]. Testing the mechanical properties of 3D-printed structures is one part of the technical research that goes into the development of AM.

One separate and important issue that is connected to additive manufacturing in a "strong" way are the intellectual property rights (IPR) related to both the printed products [2, 18, 19, 27, 28] and materials [23, 28]. As the blueprint, or "recipe", used in printing is a non-material entity that is owned in closed business models typically by the designer, it is a key component in controlling the overall AM process. Safeguarding the design and the IPR vested in designs are generally issues that may require extensive and even costly actions. In fact, IPR questions must be answered in a general and satisfactory way before additive manufacturing can make a global breakthrough.

Considering the above, one can reasonably state that the AM is not a "one size fits all"–solution, when it comes to deciding the most suitable method of manufacturing on a technical *process level*. This is why it is very interesting and important to understand, where additive manufacturing can make a difference in a business of manufacturing that is, where additive manufacturing technology-based business models make sense and what kinds of applications of additive manufacturing make business sense. Furthermore, it makes sense to understand whether there is, what can be called "transformative power" in additive manufacturing from the business model point of view. In fact, most cases such transformative power exists, because if additive manufacturing technologies make a breakthrough in some areas they will most likely also disrupt the existing *supply chains* (SC) in the related industries. This is due to the fact that the designs used in additive manufacturing are digital and can travel at the speed of light to the location at which the additive manufacturing facility resides, closest to where the product (to-be-printed) is needed. This means, for example, that printing products on location instead of moving them from one place to another [29] may become one of the ways of doing business in the logistics sector.

Revolutions in how business is done such as the one described above can, if they are widely adopted, lead to large shifts in global value chains (GVC) and may possibly obsolete, first, some of the existing ways of conducting business and, second, the infrastructure connected to it. A key difference between AM-based business models and the conventional business models is that production can be made more on-demand and the necessity to store an inventory is reduced (further). Furthermore, the production of ready-to-use products can done on-location, or nearer to location [29, 30], which means that in some cases also intermediate products do not have to be dealt with [20, 28]. All in all what one can observe is that the potential for large changes touches the logistics of manufacturing, including what is being shipped, stored, and the origin and destination of the traffic. Importantly, if significant competitive advantage can be obtained by reinventing the way business is conducted, such a change can have global implications in the overall distribution of wealth and may favor the early adopters of AM-technologies. Even though some futuristic scenarios (e.g. [15] underline the possibility of some underprivileged user-entrepreneurs printing themselves out of poverty, we tend to believe in more realistic visions provided by [31, 32] where the early adopters of AM are typically well-established large, industrial actors located in the (already) wealthy and technologically advanced societies.

Despite the clear weight and importance of looking at the effect AM has on business models and on the global value chains, it has been observed that serious existing academic literature on the topic is lacking and what is available are practitioner and consultancy reports [18, 33]. In the media the disruptive nature and effect of AM tend to be hyped, while academic literature is typically more conservative [34].

A review article can also be used in the development of fresh ideas, rather than merely concentrate on synthesizing the existing body of research [35]. In this vein, the objective here is to shed light on what has been already written on industrial additive manufacturing based business models and to identify and to shortly discuss the most promising business models and their implications on the short- and on the long-term future, where on the short-term the changes may be *incremental*, while *disruptive* changes can take place on the long-term.

This chapter goes forward by first having a look at the short-term implications of additive manufacturing and incremental business model development and then turns to discuss the long-term development and the disruptive additive manufacturing based business models. The chapter is closed with a summary and some conclusions are drawn.

2 Short-Term Implications of Additive Manufacturing and Incremental Business Model Development

A case study of hearing aid industry [36] shows, how the adaptation of AM-technologies can happen fast, when proper (profit) incentives are in place. The same study however also concludes that already having dominance over important and complementary assets such as distribution channels, customer registers, and patents, can limit the extent at which AM technologies enable new competitors to enter in the market. In other words, it is clear that in situations where additive manufacturing would cannibalize existing "good business" there is a tendency to slow down adoption, especially in firms that already enjoy a competitively advantageous position over their rivals (for extensive discussion, see, [22]). On the short-term additive manufacturing seems to serve as complimentary to the conventional manufacturing methods and replace conventional production only where it is clearly more overall cost effective technology. It can be posited that the short-term effects of additive manufacturing are case-specific and driven by company-level business drivers [37, 38].

Generally on Current Applications

The use of AM-technologies can be divided into (1) rapid prototyping; (2) rapid tooling; and (3) rapid manufacturing [23, 26, 38, 39]. Of these, the first one is routinely used in various industrial settings, as it clearly reduces both the costs and the time to market for new products [40]. In rapid tooling applications, AM-technology is used to support conventional manufacturing processes, e.g., by producing molds. Due mostly to unresolved IPR-issues, additive manufacturing activities is typically kept in-house [29, 34] and contracted AM-suppliers typically operate from centralized locations [41] instead of providing capacity on-site.

Polymer-based AM-technologies are used to produce medical or prosthetic devices [9, 36] in the industrial setting, but also to create homemade toys or household commodities by the hobbyists. The common factors for these sectors are the high customization requirements, unitary demand, and also low standardization and the relative indifference to IPR. Metal printing technologies enable a cost effective way to produce parts that are typically either expensive to produce by using conventional manufacturing methods and/or difficult to machine [17, 42]). The current defining key-feature of AM is to produce customized parts with small lot sizes. This means that AM is not likely to replace the existing, highly automated and capital-intensive, investments in mass manufacturing machinery. The future role of AM, according to [34], may be to support these mass-production investments by replacing mass-production in the production of less frequently demanded products. This line of thought can be regarded as contradictory to the ideas presented by, e.g., [20, 28] who see AM as a primus motor in the reduction of the minimum efficient scale of manufacturing.

Additive Manufacturing in Spare Parts Service

An emerging application of additive manufacturing is the production of spare parts for technically high-end industries such as the automotive and the aerospace industries [43–46]. The aerospace industry may be the single most prolific user of additive manufacturing for components at this time.

Spare parts supply in industrial applications has some distinct characteristics, which make it an especially good match with additive manufacturing technologies that offer reduced lead times to minimize supplier inventories [47], and, at the same time, extend the time OEM (Original Equipment Manufacturer) support products [24, 48]. Typically the demand for spare parts is not uniform through time and manufacturers try to minimize the spare parts inventory, while they need to be able to deliver the demanded parts quickly. Keeping either expensive equipment ready to produce more spare parts or holding large inventories for products near their time of discontinuation is very expensive and additive manufacturing offers a way out from this dilemma. The critical question to answer when making decisions about going to spare parts manufacturing with AM is whether, when, and under which conditions it is more feasible to take into use additive manufacturing over conventional manufacturing methods for spare parts. It may make sense to migrate to AM from originally conventionally manufacturing spare parts at some point, where the demand no longer supports mass-production. The AM adoption becomes an optimization problem, where profit is the target of optimization.

Today, aggregating enough demand for the relatively high-cost AM-equipment to be profitable in distributed locations remains a pivotal issue together with existing the problems that face product quality and the speed of delivery [44, 49, 50]. This observation points to the (obvious) fact that, similar to mass-production, being able to run production close to capacity is important from the point of view of profitability also for AM, and the inability to do so is an issue [49]. In a review of [51] regarding the supply chain scenarios of AM, the least considered option for additive manufacturing was the "old-school factory"-type manufacturing of products under stable demand. However, as AM equipment can produce a quite unlimited selection of different geometries there is considerably more flexibility in what can be produced and the problem of filling equipment capacity may be a smaller hurdle in the quest for profitably running the shop. Due to the need to service spare parts is often imminent, Holmström and others [44] conclude that, as a relatively slow process, additive manufacturing are never going to fully replace the policy of storing some of the most critical spare parts in the service location.

Product Service Systems (PSS)

One driving force of AM-adoption may be the emergence of service-based business models also around additive manufacturing or supported by additive manufacturing. If the user of a "machine" pays for the usable hours rather than for the machine itself, then the onus of keeping the machine in operable condition falls on the lessor. It may be beneficial to be able to tie predictive maintenance capabilities with additive manufacturing to make the machine downtime, for which no revenue can be reaped, as low as possible. This kind of thinking is very similar to the thinking behind business models that can be found, e.g., in power generation and that are tapped into by, for example, Rolls-Royce's Power selling "power by the hour" [47] and where instead of a one-off lump sum investment payment a client pays a stable stream of revenues for the power received during the years as a service purchase.

Business models that combine products and services (that can be the manufacturing as a service) are commonly referred to as Product Service Systems (PSS), we refer the interested reader to see [17, 52]. One can anticipate the rise of PSS-based business models in high-technology industries, where the complexity of equipment is constantly increasing and thus ever more secured with IPR. Matsumoto and others [17] discuss the use of additive manufacturing specifically in parts-remanufacturing and discussion about other AM-based maintenance applications can be found, e.g., in [47, 53]. One central problem that surrounds universal application of additive manufacturing to, among others, "spare parts as a service"- businesses is the fact that so far no universal standards for industrial AM platforms exist [49]. This means that, today the seemingly similar additive manufacturing set-ups are not necessarily able to produce required spare-parts and hints that agents that are able to jointly commit to a standard, or otherwise able to create an official or a de-facto standard, may be able to reap benefits over non-standardized AM manufacturers in the PSS-business over the long haul.

3 Long-Term Implications and Disruptive Business Model Development

The applications of additive manufacturing with the highest business value potential are most likely in the printing of complete parts of assembled products, or in the printing of whole products. This business falls under the rapid manufacturing genre of applications. If AM-technologies were to develop from their current niches of manufacturing into a universally accepted and applied method of manufacturing there is a chance that also a large portion of the future capacity of manufacturing has to be built based on additive manufacturing technology. Rayna and Striukova [21] envision that (B2B) customer-owned 3D-printers might become one key complementary asset for some manufacturing companies, when uncertainties around the technology are resolved. This would mean that the manufacturing for clients could happen by clients with IPR provided by the original manufacturer. This kind of thinking highlights the existing and future capabilities of market incumbents to access customer networks, see discussion in [19]. Again the issue of standardization is important, if a customer has a non-standard set of equipment then such B2B "network use" is not possible, which makes standardization as a key enabler of the disruptiveness of additive manufacturing technologies.

Rapid Manufacturing

It is proposed in the existing literature that additive manufacturing could be used in the launch phase of a product [27], something that can be dubbed "Bridge Manufacturing" referring to the phase of bridging the understanding about whether it makes sense to invest in mass-manufacturing or not. Bridge manufacturing can also be used in situations, where a new product is launched, but the production lines commissioned for mass-manufacturing are still under construction. In essence, bridge manufacturing with AM may give an edge for the existing market incumbents to quickly update their product range and supply and to speed-up the delivery process for early (often high-price paying) customers. Bridge manufacturing with AM can also serve as a valuable option to test out new designs, before making sunk and fixed investments in mass-manufacturing equipment.

There is also literature that suggests that AM is a good choice for the manufacturing of products with a stable demand [54], and literature that names AM to be suitable for conditions of declining demand [48], which is referred here as "End-of-Life Manufacturing". These, as well as bridge manufacturing, are based on the existence of a digital model for the product. For selected end-of-life components, AM-based "digitalization" of existing production may actually be relevant. In the case of new products with uncertain demand patterns it may be a good idea to design products AM-compatible directly, even though they were to be produced initially using the methods of manufacturing. This creates an option of commencing production by using additive manufacturing methods at any time. So far, AM has not been used in bridge-manufacturing, or in end-of-life manufacturing in a notable scale.

In industrial AM-systems, it is essential to ensure the purity of the used raw materials [20] due to the risk of contamination that correlates with product quality. This makes it generally infeasible to change the printed material in machines between print-jobs (even if the machines are able to print by using multiple raw materials) and suggests that the minimum size of an "all purpose" industrial grade printing facility must include a number of material-dedicated AM-machines [26]. A Delphi-study conducted among industry experts and presented in Jiang and others [19] suggests that critical parts manufactured with AM will be produced in specialized hubs to ensure quality, whereas non-critical parts can be printed also locally. It seems that print-quality is an issue that is taken rather seriously in the literature and that may affect also additive manufacturing business-models, at least initially.

Closed-Loop Manufacturing

Creating a product in one place directly from its raw materials enables a better tracking of its individual components [20]. As enticing this might be from the environmental perspective, data-sharing on material specifications is still mostly inhibited by the existing material patents [55] add that also some of the large 3DP-equipment manufacturers cherish the cartridge-sales -based business models with close filament specifications. Even though, AM reduces the direct energy and material consumption the production of filaments (e.g. metal powders) can be a major resource consumer. The overall environmental effect is therefore a huge issue that would require further, system-level, studies [18, 20, 47, 53, 56–59].

4 Conclusions and Future Directions Development

As the AM-technology is able to overcome the evident technical issues one-by-one, the focus of research interest is likely to shift towards its industrial applications. We predict that the economies of scale will take a central role in and the scenarios of small-scale, locally operating manufacturers become marginalized. The focus in business model applications will be increasingly on developing cost-effective, centralized manufacturing capabilities able to manufacture an ever widening range of high quality products on-demand (see also [34]). This development would realize some of the key benefits of AM-technology while ensuring the economies at the same time.

We anticipate that these "factory-scale" AM-facilities locate themselves near the end-customers for fast delivery and, more importantly, within fast access to global supply chains of raw materials (relatively near harbors, airports, and railway-hubs), as the manufacturing technology moving towards additive manufacturing does not make these factories of the future independent of raw materials logistics. In fact, the selection of raw materials stored on-site would likely increase rather than decrease, assuming that one would be able (and willing) to also manufacture "intermediate components" from the scratch.

Generally speaking the importance of customization, an issue highlighted in the "AM positive literature" versus the true nature of customer-needs remains an open question. Even if many products can be custom made or tailored with low extra production costs in theory, we suggest that the vast majority of customers, whether B2B or B2C, would still prefer standardized OEM-approved make-to-stock (MTS) items for increased (and promised/assured) security and warranty. Another point that supports additive manufacturing based production of MTS-items is the fact that making to stock is a way to guarantee high-enough equipment utilization rates, which are typically important for being able to insure investment payback and in large-scale operations for creating economies of scale in terms of raw material purchases.

An important and quite intuitive point to make is that in manufacturing, as elsewhere, the bottom line of financial analysis drives the actions of agents—there will not be a drive towards additive manufacturing if there is no business case. This is as true for mass production of items with AM-technologies as it is for customized small-scale production. There must be some additional featured benefit from adopting additive manufacturing that enables reaping extraordinary profits through the adoption, such as the benefits that can be derived from puncture-free long-life car tires produced with additive manufacturing en masse and that can guarantee profits on the long run. In light of the existing literature, additive manufacturing does provide a solid, new method of production that is already applied in special applications, but the high costs of production, technical constraints, and some unresolved issues with regards to IPR make additive manufacturing today (2020) unable to make a "holistic" breakthrough. However, as technical issues and intellectual property issues are resolved and as cost of production with additive manufacturing technologies are pushed down, we feel it is inevitable that additive manufacturing will have a growing place in the manufacturing systems of tomorrow by partially replacing, but more often complementing, traditional methods of manufacturing.

References

1. T. Caffrey and T. Wohlers, "An Additive Manufacturing Update," *Appl. Des.*, no. May, pp. 27–29, 2016.
2. M. Bogers, R. Hadar, and A. Bilberg, "Additive manufacturing for consumer-centric business models: Implications for supply chains in consumer goods manufacturing," *Technol. Forecast. Soc. Change*, vol. 102, pp. 225–239, 2016.
3. T. D. Ngo, A. Kashani, G. Imbalzano, K. T. Q. Nguyen, and D. Hui, "Additive manufacturing (3D printing): A review of materials, methods, applications and challenges," *Compos. Part B Eng.*, vol. 143, pp. 172–196, 2018.

4. K. V. Wong and A. Hernandez, "A Review of Additive Manufacturing," *ISRN Mech. Eng.*, pp. 1–10, 2012.
5. I. Gibson, D. W. D. W. Rosen, and B. Stucker, *Additive Manufacturing Technologies: Rapid Prototyping to Direct Digital Manufacturing*. Springer, 2009.
6. T. Wohlers and T. Gornet, "History of additive manufacturing," *Wohlers Rep. 2014*, p. 34, 2014.
7. W. Gao *et al.*, "The status, challenges, and future of additive manufacturing in engineering," *Comput. Des.*, vol. 69, pp. 65–89, Dec. 2015.
8. H.-J. Steenhuis and L. Pretorius, "Consumer additive manufacturing or 3D printing adoption: an exploratory study," *J. Manuf. Technol. Manag.*, vol. 27, no. 7, pp. 990–1012, 2016.
9. V. Petrovic, J. Vicente Haro Gonzalez, O. Jordá Ferrando, J. Delgado Gordillo, J. Ramon Blasco Puchades, and L. Portoles Grinan, "Additive layered manufacturing: Sectors of industrial application shown through case studies," *Int. J. Prod. Res.*, vol. 49, no. 4, pp. 1061–1079, 2011.
10. S. K. Rao and R. Prasad, "Impact of 5G Technologies on Industry 4.0," *Wirel. Pers. Commun.*, vol. 100, pp. 145–159, 2018.
11. R. Strange and A. Zucchella, "Industry 4.0, global value chains and international business," *Multinatl. Bus. Rev.*, vol. 25, no. 3, pp. 174–184, 2018.
12. H.-G. Kemper, P. Fettke, T. Feld, and M. Hoffmann, "Industry 4.0," *Bus. Inf. Syst. Eng.*, vol. 4, pp. 239–242, 2014.
13. D. L. M. Nascimento *et al.*, "Exploring Industry 4.0 technologies to enable circular economy practices in a manufacturing context: A business model proposal," *J. Manuf. Technol. Manag.*, vol. 30, no. 3, pp. 607–627, Nov. 2018.
14. J. Savolainen and M. Collan, "Additive manufacturing technology and business model change—a review of literature," *Addit. Manuf.*, p. 101070, 2020.
15. A. Laplume, G. C. Anzalone, and J. M. Pearce, "Open-source, self-replicating 3-D printer factory for small-business manufacturing," *Int. J. Adv. Manuf. Technol.*, vol. 85, no. 1–4, pp. 633–642, 2016.
16. P. Holzmann, R. J. Breitenecker, A. A. Soomro, and E. J. Schwarz, "User entrepreneur business models in 3D printing," *J. Manuf. Technol. Manag.*, vol. 28, no. 1, pp. 75–94, 2017.
17. M. Matsumoto, S. Yang, K. Martinsen, and Y. Kainuma, "Trends and research challenges in remanufacturing," *Int. J. Precis. Eng. Manuf. Technol.*, vol. 3, no. 1, pp. 129–142, 2016.
18. F. Hahn, S. Jensen, and S. Tanev, "Disruptive Innovation vs Disruptive Technology : The Disruptive Potential of the Value Propositions of 3D Printing Technology Startups," *Technol. Innov. Manag. Rev.*, no. December, pp. 27–36, 2014.
19. R. Jiang, R. Kleer, and F. T. Piller, "Predicting the future of additive manufacturing: A Delphi study on economic and societal implications of 3D printing for 2030," *Technol. Forecast. Soc. Change*, vol. 117, pp. 84–97, 2017.

20. M. Despeisse *et al.*, "Unlocking value for a circular economy through 3D printing: A research agenda," *Technol. Forecast. Soc. Change*, vol. 115, pp. 75–84, 2017.
21. T. Rayna and L. Striukova, "From rapid prototyping to home fabrication: How 3D printing is changing business model innovation," *Technol. Forecast. Soc. Change*, vol. 102, pp. 214–224, 2016.
22. C. Weller, R. Kleer, and F. T. Piller, "Economic implications of 3D printing: Market structure models in light of additive manufacturing revisited," *Int. J. Prod. Econ.*, vol. 164, pp. 43–56, 2015.
23. M. Khorram Niaki and F. Nonino, "Impact of additive manufacturing on business competitiveness: a multiple case study," *J. Manuf. Technol. Manag.*, vol. 28, no. 1, pp. 56–74, 2017.
24. G. Manogharan, R. A. Wysk, and O. L. A. Harrysson, "Additive manufacturing-integrated hybrid manufacturing and subtractive processes: Economic model and analysis," *Int. J. Comput. Integr. Manuf.*, vol. 29, no. 5, pp. 473–488, 2016.
25. M. Baumers, P. Dickens, C. Tuck, and R. Hague, "The cost of additive manufacturing: Machine productivity, economies of scale and technology-push," *Technol. Forecast. Soc. Change*, vol. 102, pp. 193–201, 2016.
26. Wohlers, "Additive Manufacturing Technology Roadmap for Australia," 2011.
27. B. Berman, "3-D printing: The new industrial revolution," *Bus. Horiz.*, vol. 55, no. 2, pp. 155–162, 2012.
28. A. Laplume, B. Petersen, and J. M. Pearce, "Global value chains from a 3D printing perspective," *J. Int. Bus. Stud.*, vol. 47, no. 5, pp. 595–609, 2016.
29. M. Rehnberg and S. Ponte, "From smiling to smirking? 3D printing, upgrading and the restructuring of global value chains," *Glob. Networks*, vol. 18, no. 1, pp. 57–80, 2018.
30. B. P. Conner *et al.*, "Making sense of 3-D printing: Creating a map of additive manufacturing products and services," *Addit. Manuf.*, vol. 1–4, pp. 64–76, Oct. 2014.
31. D. R. Gress and R. V. Kalafsky, "Geographies of production in 3D: Theoretical and research implications stemming from additive manufacturing," *Geoforum*, vol. 60, no. February, pp. 43–52, 2015.
32. F. Matos and C. Jacinto, "Additive manufacturing technology: mapping social impacts," *J. Manuf. Technol. Manag.*, vol. 30, no. 1, pp. 70–97, Aug 2018.
33. J. Gartner, D. Maresch, and M. Fink, "The Potential of Additive Manufacturing for Technology Entrepreneurship: An Integrative Technology Assessment," *Creat. Innov. Manag.*, vol. 24, no. 4, pp. 585–600, 2015.
34. A. Sasson and J. C. Johnson, "The 3D printing order: variability, supercenters and supply chain reconfigurations," *Int. J. Phys. Distrib. Logist. Manag.*, vol. 46, no. 1, pp. 82–94, 2016.
35. J. Webster and R. T. Watson, "Analyzing the Past to Prepare for the Future: Writing a Literature Review," *MIS Q.*, vol. 26, no. 2, pp. xxiii–xxiiii, 2002.

36. C. G. Sandström, "The non-disruptive emergence of an ecosystem for 3D Printing—Insights from the hearing aid industry's transition 1989–2008," *Technol. Forecast. Soc. Change*, vol. 102, pp. 160–168, 2016.
37. S. Mellor, L. Hao, and D. Zhang, "Additive manufacturing: A framework for implementation," *Int. J. Prod. Econ.*, vol. 149, pp. 194–201, 2014.
38. D. G. Schniederjans, "Adoption of 3D-printing technologies in manufacturing: A survey analysis," *Int. J. Prod. Econ.*, vol. 183, no. October 2016, pp. 287–298, 2017.
39. C. Achillas, D. Aidonis, E. Iakovou, M. Thymianidis, and D. Tzetzis, "A methodological framework for the inclusion of modern additive manufacturing into the production portfolio of a focused factory," *J. Manuf. Syst.*, vol. 37, pp. 328–339, 2015.
40. M. Attaran, "The rise of 3-D printing: The advantages of additive manufacturing over traditional manufacturing," *Bus. Horiz.*, vol. 60, no. 5, pp. 677–688, 2017.
41. H. Rogers, N. Baricz, and K. S. Pawar, "3D printing services: classification, supply chain implications and research agenda," *Int. J. Phys. Distrib. Logist. Manag.*, vol. 46, no. 10, pp. 886–907, 2016.
42. M. Kathryn *et al.*, "CIRP Annals—Manufacturing Technology Design for Additive Manufacturing : Trends, opportunities, considerations, and constraints," *CIRP Ann.—Manuf. Technol.*, vol. 65, no. 2, pp. 737–760, 2016.
43. S. H. Khajavi, J. Partanen, and J. Holmström, "Additive manufacturing in the spare parts supply chain," *Comput. Ind.*, vol. 65, pp. 50–63, 2014.
44. J. Holmström, J. Partanen, J. Tuomi, and M. Walter, "Rapid manufacturing in the spare parts supply chain: Alternative approaches to capacity deployment," *J. Manuf. Technol. Manag.*, vol. 21, no. 6, pp. 687–697, 2010.
45. L. F. C. S. Durão, A. Christ, R. Anderl, K. Schützer, and E. Zancul, "Distributed Manufacturing of Spare Parts Based on Additive Manufacturing: Use Cases and Technical Aspects," *Procedia CIRP*, vol. 57, no. January 2017, pp. 704–709, 2016.
46. M. Savastano, C. Amendola, F. D'Ascenzo, and E. Massaroni, "3-D Printing in the Spare Parts Supply Chain: An Explorative Study in the Automotive Industry," in *Lecture Notes in Information Systems and Organisation*, vol. 18, 2016.
47. S. Ford and M. Despeisse, "Additive manufacturing and sustainability: an exploratory study of the advantages and challenges," *J. Clean. Prod.*, vol. 137, pp. 1573–1587, 2016.
48. M. Gebler, A. J. M. Schoot Uiterkamp, and C. Visser, "A global sustainability perspective on 3D printing technologies," *Energy Policy*, vol. 74, no. C, pp. 158–167, 2014.
49. J. Holmström, J. Partanen, and J. Holmström, "Digital manufacturing-driven transformations of service supply chains for complex products," *Supply Chain Manag.*, vol. 19, no. 4, pp. 421–430, 2014.

50. D. R. Eyers and A. T. Potter, "E-commerce channels for additive manufacturing: an exploratory study," *J. Manuf. Technol. Manag.*, vol. 26, no. 3, pp. 390–411, 2015.
51. M. J. Ryan *et al.*, "3D printing the future : scenarios for supply chains reviewed," *Int. J. Phys. Distrib. Logist. Manag.*, vol. 47, no. 10, pp. 992–1014, 2017.
52. C. Vezzoli, F. Ceschin, J. C. Diehl, and C. Kohtala, "New design challenges to widely implement 'Sustainable Product–Service Systems,'" *J. Clean. Prod.*, vol. 97, pp. 1–12, 2015.
53. M. Despeisse, M. Yang, S. Evans, S. Ford, and T. Minshall, "Sustainable Value Roadmapping Framework for Additive Manufacturing," *Procedia CIRP*, vol. 61, pp. 594–599, 2017.
54. J. Minguella-Canela *et al.*, "Comparison of production strategies and degree of postponement when incorporating additive manufacturing to product supply chains," *Procedia Manuf.*, vol. 13, pp. 754–761, 2017.
55. A. Garmulewicz, M. Holweg, H. Veldhuis, and A. Yang, "Disruptive Technology as an Enabler of the Circular Economy: What Potential Does 3D Printing Hold," *Calif. Manage. Rev.*, vol. 60, no. 3, pp. 112–132, 2018.
56. J. Faludi, C. Bayley, S. Bhogal, and M. Iribarne, "Comparing environmental impacts of additive manufacturing vs traditional machining via life-cycle assessment," *Rapid Prototyp. J.*, vol. 21, no. 1, pp. 14–33, 2015.
57. S. Hankammer and R. Kleer, "Degrowth and collaborative value creation: Reflections on concepts and technologies," *J. Clean. Prod.*, vol. 197, pp. 1711–1718, 2018.
58. S. H. Huang, P. Liu, A. Mokasdar, and L. Hou, "Additive manufacturing and its societal impact: A literature review," *Int. J. Adv. Manuf. Technol.*, vol. 67, no. 5–8, pp. 1191–1203, 2013.
59. M. A. Kreiger, M. L. Mulder, A. G. Glover, and J. M. Pearce, "Life cycle analysis of distributed recycling of post-consumer high density polyethylene for 3-D printing filament," *J. Clean. Prod.*, vol. 70, pp. 90–96, May 2014.

Additive Manufacturing Cases and a Vision for a Predictive Analytics and Additive Manufacturing Based Maintenance Business Model

Michele Urbani and Mikael Collan

1 Introduction

In the previous chapter we have seen that the literature on additive manufacturing business models can in broad strokes be divided into four different directions. To illustrate the real-world status quo with examples we discuss in this chapter two real-world cases in the context of using additive manufacturing technology in the production of medical prostheses and in the refurbishment of metal dies and discuss the business model aspects of both of these cases. The third part of the chapter is used to

M. Urbani
Department of Industrial Engineering, University of Trento, Trento, Italy
e-mail: michele.urbani@unitn.it

M. Collan (✉)
LUT University, Lappeenranta, Finland
e-mail: mikael.collan@lut.fi

M. Collan, K.-E. Michelsen (eds.), *Technical, Economic and Societal Effects of Manufacturing 4.0*,
https://doi.org/10.1007/978-3-030-46103-4_7

discuss a more visionary case, where additive manufacturing, together with predictive maintenance, allows one to rethink how the modern system of maintenance could work.

2 Additive Manufacturing Used in Enhancing Heart Surgery

The use of additive manufacturing in healthcare applications has flourished in the past two decades [1] and the market share of additive manufacturing in this market was $6.1 billion in 2016 [2]. The market share of additive manufacturing has caught a steady uptrend, a dramatic increase towards $21 billion in 2020 is expected.

The improvement of old and the creation of new techniques for 3D printing, together with the development of new purpose-specific materials for the healthcare sector have made possible the deployment of a range of patient-specific applications [3]. For instance, the customization of healthcare products and services such as the realization of customized prosthesis, and the development of case-sized in-vitro models would not have, in many cases, been possible without the developments made in additive manufacturing technologies.

In this chapter we discuss the use of additive manufacturing in the treatment of heart disease from the points of view of the medical procedure involved and the technical solution that additive manufacturing can offer. Cardiovascular diseases (CVD), which are the focus of this real-world example, are a top ranked cause of death on a global level. All in all they were responsible for 17.9 million of deaths in 2015, with almost a 50% increase from 1990 [4]. The added value of additive manufacturing in the process is discussed. The chapter is based on the real-world collaboration between the Department of Industrial Engineering of the University of Trento (Italy) and the Azienda Sanitaria per i Servizi Sanitari (public company in charge of the provision of healthcare services) of the autonomous Province of Trento (Italy).

Atrial Fibrillation: The Condition and the Surgical Intervention

Persons who suffer from a condition known as the non-valvular atrial fibrillation (AF) may be subject to the occurrence of blood clots, which after being formed within the left atrial appendage (LAA) can enter the blood stream and cause a stroke, or other vascular complications. Many patients are regularly treated with oral anticoagulants, which are aimed at preventing the formation of the blood clots. Unfortunately, this therapy is not always possible, since some individuals have low tolerance of anticoagulants or the risk of bleeding problems caused by the anticoagulants is too high.

An alternative to using anticoagulants is to permanently seal off the LAA—the procedure does not solve the problem with AF, but it helps to prevent blood clot formation through the closure of the LAA.

The surgical intervention in question is called *left atrial appendage occlusion*, also known as the left atrial appendage closure. It is a non-invasive non open-body surgical intervention. There are a number of techniques that can be used to occlude the LAA, one of the most recently introduced practices consists of placing an implant via trans-esophageal echo guidance. The purpose of the implant is to ensure the closure of the LAA and to impede the flow of blood. The intervention is carried out under general anesthesia and is similar to the implantation of a stent.

Since the geometry of the human heart is different for each patient, the *size* and *shape* of the implants to be installed are of fundamental importance. In this context, the decision-maker is the surgeon, who bears the responsibility for the outcome of the surgery. The standard process to treat atrial appendage occlusion begins with a computed tomography (CT) scan of a patient's chest. This allows the doctor to create a rough estimate of the shape and size of the implants that will be implanted. CT is an effective tool and provides a set of cross sectional images of the scanned body along the sagittal plane. The set of images can be processed via a 3D-software and a model of the heart can be created—this allows the patient's LAA to be inspected along the desired direction and gives the surgeon a better basis for decision-making. While the CT scan images and the 3D model are helpful, it remains difficult to foresee the practical difficulties that may arise during the operation.

In the current practice, implants are mass-produced according to standardized shapes and sizes, which forces the surgeon to choose from among a set of possible implant alternatives. With the aid of the CT scan, the doctor can narrow down the set of implants that could fit the heart of a given patient, but there is no full-proof way to in advance identify the right implant alternative.

In practice the fit of the implant is directly tested on the patient during the surgery. Once the right dimensions have been found and the final decision on the implant made, the implant is implanted. The regular procedure foresees that patients can be released after a 24 hour recovery which is typically followed by a 45 day anticoagulant treatment. The success of the intervention cannot be determined immediately after the execution, due to the time required by the human body to adapt to the presence of the implant, in fact the verification takes place during the weeks following the surgery through periodic checks

In case the procedure fails, the operation is typically repeated and the implant is replaced with a better fitting one, thus subjecting the patient to a second intervention. The failure of the process may have severe consequences for the patient, such as pericardial effusion, incomplete LAA closure, dislodgement of the device, and other risks related to catheter-based techniques [5].

Enhancing the Procedure with the Help of Additive Manufacturing Technology

Additive manufacturing can be used to reduce the uncertainty connected to making the selection of the implant used and to assist in planning the surgery ex-ante. Contrary to what the reader might expect, the target of using additive manufacturing in this context is not the creation of a tailor-made implant, but finding a best fitting mass produced implant. While it is logical to expect that once a 3D model of the patient´s heart is available an implant of the right size and shape could be additively manufactured, however the low cost of the mass-produced implant and their availability on the market does not make it profitable to print them. Instead a real-size copy of the patient´s heart is printed, based on the 3D model obtained from the CT scan. This allows the surgeon to test fit of the (different) mass-produced implants. The process is relatively low cost and provides the advantage of being able to perform pre-surgery testing of fit and by reducing uncertainty increases the chances of a successful operation. The life-size printed model of the heart also allows the surgeon to study the heart and the execution without any pressure, which may prevent practical difficulties during surgery to become overwhelming.

The 3D model produced from the CT images is practically print-ready, no processing by a human operator is needed. A software tool provided by the 3D printer supplier analyses the model and schedules the work for the printer and adds the print-support structures needed, before the model can be sent to the machine for printing. The printing technology used is stereolitography and it is one of the oldest 3D printing technologies. Patented in the United States in 1986, stereolitography uses a generic photopolymer—specifically a thermosetting resin monomer to build the printed object layer-by-layer on a build platform. Each layer is solidified by a UV laser beam that moves all over the cross-section and is used to solidify the resin. Once a layer has been completed the build platform is lowered and a new layer of material is injected on the cross section—before the new layer is solid, the excess liquid resin is wiped out in order to obtain the right layer-thickness; this also ensures that the surface is even enough for the application of the forthcoming layers. Finally, the printed heart must be finalized by removing the printing supports mechanically (by hand).

The material used (resin) does not have to be bio-compatible as it is not used in contact with the human organs and it can be chosen based on needed mechanical properties of which elasticity is the most important in being able to replicate human tissue-like behaviour during the testing and surgery planning phase. The resin typically used is the softest provided by the supplier and has a Shore-hardness of 50A. Compared to many other photopolymers for 3D printing, the substance is very elastic and it can reach 160% of elongation before breaking. The printing process including the post-treatment of the polymer takes a few hours depending on the complexity of the printed object. The short printing-time together with a good surface finish of the end product, print resolution is up to 100 microns, make stereolitography a suitable technology for this application.

The Business Model Perspective

The application of additive manufacturing described above presents features that are also found in previous literature [6] as hallmarks of successful implementation of additive manufacturing in practice—low production volume, customization of the product, on-demand production, the availability of the (3D) model, and the (modest) cost of the printing equipment.

The cost of a suitable stereolitographic printer ranges between €3000 and €5000, which in the context of the healthcare sector is a rather affordable investment for most medium and large hospitals, if knowledgeable personnel is already in place. The cost of the resin monomer is approximately 200 €/l, which translates to a material cost of some tens of euros per printed heart. In the context of this case, the region of Trentino—Alto Adige in the North of Italy, the yearly number of operations of the described type is less than one hundred; the region has approximately one million inhabitants.

The low volume of printed hearts produced may nevertheless make it unreasonable for a single hospital to acquire a printer, in which case renting the printing capacity could be a more cost-effective solution. Many research-centres and universities are likely to own a suitable printer and a partnership between hospitals and local research-institutes are a logical way to create win-win partnerships around this theme. More importantly, research-institutes typically have the manpower and expertise to manage the printing process. Starting from a scratch, the time required to acquire the knowledge to manage a polymeric printing process is reasonable—with a few weeks of training, a person is able to manage the whole process.

To summarize, in the present context it does not make sense to talk about a proper business model for this additive manufacturing application, however there is a clear benefit to using additive manufacturing to enhance the treatment of atrial fibrillation. On a scale of multiple hospitals and some thousands of printed hearts on an annual basis there might be a profitable niche for a specialized AM manufacturer. There is always business relevance in being able to provide superior techniques for medical purposes that lower the risks of surgical interventions—this refers to what has previously been characterized as incremental change in adopting additive manufacturing [7].

3 Refurbishing Metal Dies with 3D-Printing

In the context of manufacturing, it is very common to have processes that require a physical contact between a manufacturing machine and the processed product. When the contact is made in order to modify the shape of the processed item it is inevitable that the part of the machine that makes the contact, the so called *die*, will be subject to wear. The die can be made of different materials, but here we concentrate on metal dies. Many different surface treatments and improvements in the grade of the base material for dies have been designed in order to limit the effects of wear on metallic dies. While the metals used are hard, on the long run it is inevitable that a die looses nominal geometry, or the surface of the die becomes defective.

The more severe defects are typically surface cracks, sub-surface cracks, and the loss of material from the surface of the die. Generic loss of the nominal geometry is a type example of less severe defects that occurs with time in the most highly stressed areas of the die. Change of die geometry is a common cause of low quality in the end product. The general picture is that wear causes the end product of the process involving the die to decrease in quality and at a point when the quality-decrease at a limit to acceptable level the die used must be replaced or refurbished.

Due to the above issues, some metallic parts of manufacturing systems must in practice constantly be monitored through the inspection of the quality of the finished items, or through the inspection of the status of the parts themselves. Control performed by way of inspecting finished items is typically based on comparing the produced parts to the design specifications and when the design tolerances are no longer satisfied action must be taken. There are opportunities for predictive maintenance in these cases—minimizing the number of unsatisficing end products is a cost issue.

Maintenance of a system component most often requires stopping the machine, which causes a loss in productivity—this is why the need to repair machines quickly and efficiently is as old as the manufacturing industry itself. Here we concentrate on the maintenance of metal dies used in manufacturing and especially on the refurbishing of metal dies by way of additive manufacturing technologies.

Refurbishing Metal Dies

The current practice of refurbishing metallic dies is based on manual labour. After the defective die has been found via a visual inspection of the production line, it is removed and prepared for maintenance. If the defect is a surface-crack the damaged region of the die will typically consist of an irregular surface on which work cannot be done—machining is first done to remove irregular surfaces, this is done by an operator by hand with a milling machine. In this task the die must be carefully placed in the milling machine and the position of the die and the machine must be calibrated. The end-result is a cavity with a smooth surface.

The cavity is then filled with a suitable filler-metal, typically a manual electro-welding process is used. After filling the cavity the die undergoes re-machining so that the original required (nominal) geometry is re-obtained. This means that a milling machine is again used, after loading the die and calibration of the position on the machine. An error in the positioning of the die in the milling machine will compromise the success of the whole operation. While the manual refurbishing of the die is relatively inexpensive the risks related to the positioning and calibration of the die in the milling machine remains a problem. The main phases of die-refurbishment are visible in Fig. 1.

The Hybrid Manufacturing Approach to Refurbishing Metal Dies

Thanks to recent developments in the field of manufacturing equipment development, new *hybrid* equipment has become available. A hybrid manufacturing workstation embeds two, or more, manufacturing technologies within it. Typically this means that the elements of ***both additive and subtractive manufacturing are present in the same system***. The clear advantage of a hybrid workstation is that as it is able to perform a large number of operations the set-up costs are typically lower. Specifically, only one instance of pre-processing (including calibration) is needed if the hybrid workstation is able to perform an operation, for which multiple machines are otherwise needed—this may dramatically reduce the time consumption as well as the risks related to pre-processing. Hybrid work stations are operated by software designed specifically for these machines. The downside of modern hybrid machines is their relatively high cost.

The hybrid workstation used in refurbishing dies is a DMG Mori Lasertec 65—the workstation integrates laser-deposition melting technology with a 5-axis milling station. The station is able to automatically change between the laser- and the milling-heads. Limitations that the workstation has have to do with the volume and the weight of the worked-on parts (Ø 500 mm × 400 mm; 600 kg)—this kind of limitations are "real" in terms of the workstation not being able to handle larger and heavier objects; as technology is developed further these limitations are slowly relaxed, but the limitations mentioned are on a "good modern

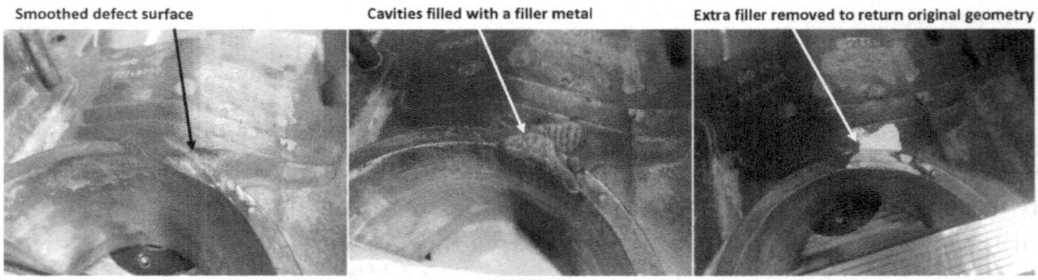

Fig. 1 The three main phases of refurbishing metal dies. (Original photos Matteo Perini)

level". In heavy industry the die component size can still be too large to fit into hybrid workstations for quite a while. The workstation is able to handle various metals and alloys that include stainless steel, nickel-based alloys (Inconel 625, 718), tungsten carbide matrix materials, bronze and brass alloys, chrome-cobalt-molybdenum alloys, stellite, and tool-steel. The CAM/CAD software used is the Siemens NX.

The process of refurbishing dies with the hybrid workstation begins with the setting up and calibration of the damaged die in the workstation and is followed by a 3D-scan and the subsequent construction of a virtual model of the damaged die. A separate software is used for 3D-scanning. The accuracy of the virtual model depends on the resolution of the scanner. The accuracy is a relevant issue, as the more accurate the model is the more accurately it can be decided, which parts of the damaged areas need to be removed—typically the more is removed the more needs to be added later on. If the metal alloy used is very expensive the ability to use less material may have a positive effect on the total cost.

After the decision has been taken, the virtual model is compared to a model of the original (nominal) geometry of the die. With the original model and the virtual model of the damaged die it is possible to obtain the difference between the two and "instruct" the workstation to reconstruct the nominal geometry. The accurate reproduction of the original topology by additively filling the cavities to be repaired is the result of a focused research project at the University of Trento that developed a new method [8] and supporting software that translates the topological difference to a set of machine-understandable instructions that the CAM software is able to read.

The laser deposition melting solution used allows a homogenous distribution of metallic powders, which occurs under the protection of a shield gas that protects the process from oxidation. The system construct is such that a separate work chamber with a controlled environment is not needed—this makes the process faster that is typically the case. It must be observed that the die typically consists of two metallic parts—the part that can be called a "saddle" that is connected to the machine and the "contact part" that is made of a harder metal and that is attached to the saddle and that is the part of the die that is in contact with the produced parts. The actual additive manufacturing procedure is divided into three layers, where the first layer is the (material of the saddle part of the) refurbished die, the second layer is called a dilution zone and it is a mixed material made partly of the original die saddle metal and partly of the filler metal (contact part), and the third layer is fully made of the filler metal. This three-part procedure is able to produce a very durable *non-porous* and *crack-free* metallic solid—the refurbished die can be said to be "as good as new", which is the best possible end-result.

The Business Model Perspective

Similar to the heart implant example above, also in the case of die refurbishing the uniqueness of the procedure and the product are key elements—that is, the unique faults in the dies offer a possibility for additive manufacturing to be competitive. Furthermore, as the dies are typically constructed of two metals the reconstruction process of a die is not simple and the ability to refurbish dies to "as good as new" state requires handling high product complexity in an efficient way, something that is possible with the hybrid workstations presented above. This also means that if dies are refurbished en masse that there is always an element of customization to the work—identifying the refurbishing procedure for the various kinds of faults allows something that resembles mass customization. If a relatively expensive hybrid workstation is acquired with a profit in mind it is clear that the workstation should have an as high as possible rate of utilization. This means that there should be a number of different dies (and other parts) for the refurbishing of which the processes should be well-known and ready.

In such a case, even a single hybrid workstation could act as a part of a number of maintenance supply chains and in essence function as a machine-as-a-service (MaaS). The workstation could be purchased through a leasing contract by the customers, who pay an annual fee for the use of the machine, or the machine is acquired by one "player" who then sells the capacity of the machine to others—there are many possible types of ways to organize the availability and the sale of the capacity of the workstation. In the case of the refurbishing metal dies the workstation can, e.g., be a part of a die maintenance chain that consists of predictive maintenance system in place at one or several manufacturing facilities that use(s) metal dies and that is able to refurbish-on-demand. Persona et al. [5] write about maintenance outsourcing and the resulting effects on supply chain organization.

For a manufacturing company the number of dies that need refurbishing on recurring basis must be large enough to warrant the relatively high costs of acquiring and operating a working station, which indicates that such a move would make sense only for large-enough operations. If a workstation is present on-site any logistics costs are reduced—this may have a marginal positive effect on the cost side, however the potential to shorten downtimes with on-site refurbishing may have a more remarkable positive effect. In a broader perspective the adoption of additive manufacturing technologies must be regarded as a strategic choice for a firm. Purchasing AM capable machinery, such as the hybrid workstation, an organization makes a long-term commitment into a new technology, which not only includes the cost of equipment, but demands the acquisition of the related human talent. We refer the interested reader to Weller et al. [9] for additive manufacturing cases in maintenance applications.

One option in this space is to outsource the maintenance of the metal dies and buy "dies as a service". There are specialized firms that exclusively sell industrial maintenance capacity and in a sense machine availability—typically in these cases the production facilities belong to customer (manufacturing company) and the service provider is in charge of their good functioning. This option will be discussed more in the remaining part of this chapter, where disruptive maintenance-related business models that rely on digitalization and excellence in additive manufacturing are presented.

4 Predictive Maintenance and Additive Manufacturing: Joint Business Model

Broadly speaking, predictive maintenance is the practice of scheduling and performing maintenance in a way that predicts failures and is hence able to contribute to minimizing production downtimes, maximizing component lifetimes, and to minimizing maintenance costs, we refer the interested reader to see [10]. The indirect benefits that predictive maintenance brings include the potential to use maintenance resources more efficiently, the ability to carry a lower inventory of spare parts, and the important ability to make "tougher" production-related promises to customers. These benefits accrue to both the owner of the maintained system and to the organization responsible for the maintenance that can also be the same organization. The ability to make maintenance more efficient is a source of lasting competitive advantage.

Predictive maintenance is winning ground in manufacturing (and elsewhere) due to the instrumentation of manufacturing equipment that allows automating the collection of condition data. Based on the data collected predictive models can be tuned in a way that enables the accurate prediction of the timing of equipment failures and the construction of smart maintenance schedules.

Different architectures for predictive systems exist, perhaps the most prevalent at the time of writing are "monitoring-based" systems that track deviations in the system captured by sensors and alert as they appear. Different types of deviations may have different types of "fingerprints" and known the tell-tale signs of a deviation allows the correct classification of the deviation and the correct prediction of an incoming fault. These systems are evolving in the sense that their ability to identify failures becomes better with time as more and more data is accumulated and the patterns that distinguish the different failures become better known. In essence these systems utilize "machine learning".

Smart means in this context also the ability select a good maintenance policy that keeps the level of unexpected component failures (and stoppages) at an acceptable level. Smart increasingly means also being able to answer to more difficult questions such as: "once equipment is shut down for maintenance, what else than only the minimum necessary maintenance should one do?"—questions of this type and finding answers to them is difficult and requires system-size modeling for maintenance optimization.

Maintenance optimization work typically includes the modeling of the maintained system and the individual maintainable components (including modeling the wear and tear) and the optimization of the system maintenance based on the model. Bundling maintenance actions in an optimal way is a complex optimization problem and requires considerable computing power and good modeling. So far the typical target of maintenance optimization has been a single system, however, it is clear that the optimization of multiple systems simultaneously offers added benefits. If issues such as workforce scheduling are also taken into consideration in the optimization the complexity of the optimizable problems increases, but on the other hand so do the potential rewards.

One can without a doubt make the claim that the sophistication needed in maintenance optimization is at par with the sophistication needed in the rest of the Manufacturing 4.0 paradigm—someone might even go as far as to say that smart maintenance can be seen as a part of the paradigm when the maintenance context is manufacturing.

Predictive Maintenance Based Business Model for Additive Manufacturing

What makes predictive maintenance different from "typical maintenance" is that due to the instrumentation in place even ad hoc (un-expected) failures can be predicted—in other words there is typically sufficient time to react between acknowledging that a component is about to malfunction and the actual time the component breaks. This period of grace that results from the predictability of faults can be utilized to render the manufacturing operation more efficient by determining the optimal maintenance actions that are performed, when the component that is known to malfunction is changed and by making ready the preparations for the said actions to be performed—including procuring the needed components that need to be exchanged. In this context the procuring the components is the key issue, because the new replacement components can be taken from a (local) storage if they are available, brought to the failing machine from a storage or production location further away, or produced on-site (or near) by additive manufacturing.

Implications of enforcing and making stronger the connection between predictive maintenance and additive manufacturing are quite remarkable—in cases where a failing component, or a spare-part, can be manufactured in time and on-demand for the maintenance action to take place, there may be and there most likely are savings to be made. In the case, where the alternative is transporting a spare-part from far away, which is by far not unheard of. If the on-site (additive) manufacturing of spare parts becomes the trend, the logistics of spare-parts becomes less of an issue and in fact the "logistics middleman" can be even completely cut out. Spare parts logistics are replaced by the logistics of the much less expensive and non time-critical logistics of the materials needed to produce the spare part(s) on location and the digital logistics of the information needed to print the spare part.

One can observe that also the need for the storage of spare parts is diminished as only spares that cannot be printed on demand must be stored—as time passes it can be expected that the selection of materials available for additive manufacturing grows wider and the quantity of non-printable parts grows smaller. Generalizing and perhaps being slightly polemic one may surmise that if there is a revolution by additive manufacturing, then it surely must also be a revolution in logistics. As logistics costs are not insignificant there is a clear potential for savings immediately, when the cost of production of spare parts by way of additive manufacturing become competitive. We refer the reader interested in the supply chain effects of additive manufacturing to see [11].

The above described predictive maintenance—additive manufacturing symbiosis requires quite seamless informational collaboration between the activities of maintenance (and operation of the maintained equipment), which typically require full knowledge of the design of the said machine, and of the parts production for the machine. In other words, the collaboration of a number of stakeholders in the process is necessary in a way that is very fast, and in the best case automatic.

Automation means that there is a need for a standard system level "rules of play" that govern the informational and trade exchanges taking place within the system, including a joint understanding and pre-acceptance of the involved costs. With the costs we refer, among other things, to the cost of the rights of use of the "recipe" or the digital plans required to print the spare parts, whose IPR typically resides with the original equipment manufacturer.

The fact that a number of things need to be pre-planned and pre-accepted creates a great a "natural" hurdle, when (multiple) separate organizations need to reach consensus—it is therefore likely that the first working systemic solutions that incorporate these technologies in the way envisioned above are formed by actors that already control the different steps of the maintenance and spare-parts production whole and are therefore able to benefit from any and all efficiency increases and cost savings related to process changes.

Blueprint for a Vision

Instrumented equipment is able to digitally transmit real-time information about the condition of perishable parts to what is called "predictive maintenance optimization system" in Fig. 2. The idea is that a sophisticated maintenance analytics system is able to utilize data coming from the sensors located in the production equipment (#1 in Fig. 2), to create results by utilizing modern condition-based maintenance and predictive maintenance models (for ad-hoc failures), and to use the results as input in a sophisticated maintenance action optimization. Modern optimization systems are able to intelligently group maintenance actions to realize potential cost savings from performing multiple maintenance actions simultaneously. Putting smart maintenance planning automatically into

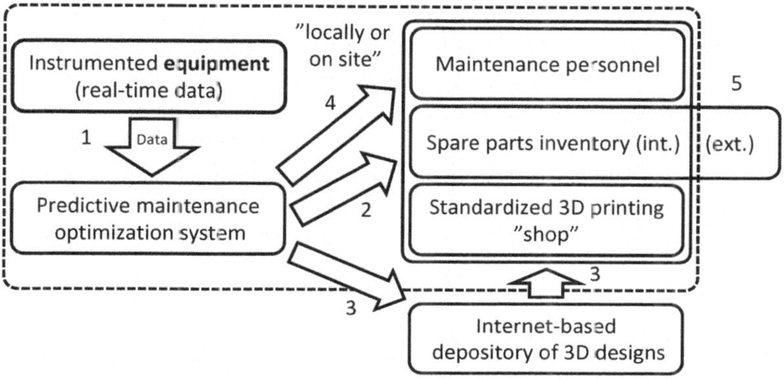

Fig. 2 Additive manufacturing "3D printing shop" as a part of a predictive maintenance oriented maintenance ecosystem

action means that the optimization system has the ability to check the local spare parts inventory for existing parts that are needed (#2 in Fig. 2), to order needed designs from the on-line OEM depository of 3D designs for the parts that need to be printed, and reserving time slots for printing from the 3D printing facility (#3 in Fig. 2).

The maintenance optimization system can also reserve (and in some cases even optimize) the maintenance personnel resources needed and schedule the actual maintenance (#4 in Fig. 2). In contrast to the traditional model, where the locally non-available parts would be searched for and ordered by the automatic system and then shipped to the location from an external warehouse possibly on another continent (#5 in Fig. 2) the additive manufacturing based model can allow for all physical actions to be performed on location. It is clear that a hybrid of the "old and the new" is a state that may be in place for a long time and where the smart optimization system is ultimately able to decide whether to order spare parts from an external warehouse or from a local 3D-printing facility based on minimizing a cost function that may include, e.g., time penalties.

In an ideal world the optimization system is able to create a circumstance, where the costs are minimized, optimal amount of maintenance is carried out, parts are ready just-in-time, and personnel resources are optimized. The driving forces behind reaching this kind of a state are the development of digital instrumentation in equipment (IoT), development of smart predictive maintenance systems that are coupled with advanced maintenance optimization systems that are digitally connected to resource management systems. The vision presented includes an Internet-based depository of 3D designs as a component—such depositories already exist for hobbyist designs, as of yet serious B2B depositories have not emerged.

Many different kinds of business model possibilities exist within the vision, the envisioned whole can be realized within the "realm" of a single actor, or by way of collaboration of specialized single actors.

5 Conclusions

This chapter has concentrated on presenting two real-world cases of how additive manufacturing can be used to enhance existing processes that otherwise demand precision manual labor and/or cannot be performed as well. Both of the real-world cases show that there is potential in additive manufacturing in places, where sophisticated tailoring of what is done is required, and where precision is a key factor. In both the cases the business model aspects had not been fully explored due to the exploratory and piloting nature of the activities performed, but it remains quite clear that with a high-enough demand for the presented activities there is a profitable business case to be made. If a specialized know-how is created around an additive manufacturing resource, the resource can be leveraged to service multiple different clients. It must be observed that in the same way as with any production technology, if the utilization rate of the equipment used is low the chances of reaching profitability remain a challenge—the laws of production economics do not change.

There seems to be a place for visioning additive manufacturing based business models that combine additive manufacturing with other technologies, such as predictive maintenance, as presented in this chapter. The ability of additive manufacturing to deliver on-demand is an important factor from the point of view of efficiency gains it is able to bring to the business of which additive manufacturing is a part of. When coupled with "control" technology that is able to make just-in-time orders and to optimize processes the ability to produce just-in-time can be exploited effectively. The prospect of locally manufacturing with additive manufacturing technologies through a global web of digital information is an interesting one and puts pressure on mass-production and long-haul logistics based business models.

Industry-grade 3D-printers can be thought of as platform investments that service more than on client and that draw from a world-wide resource of 3D-printing designs. At this time serious commercial business to business depositories of 3D-printing designs do not exist and the business model is still in its infancy. Many issues remain to be solved in the (digital) collaboration between the original equipment manufacturers to whom the 3D-printing designs belong to, the secure distribution and pricing of designs, and the (trusted) network of 3D-printing resources that can service clients globally.

References

1. S. H. Huang, et al., "Additive manufacturing and its societal impact: a literature review," *International journal of advanced manufacturing technology*, vol. 67, pp. 1191–1203, 2013.
2. I. Campbell, et al., "Wohlers Report 2017 – Additive Manufacturing and 3D Printing State of the Industry Annual Worldwide Progress Report," Wohlers Associates Inc. , Fort Collins, Colorado 2017.
3. A. A. Zadpoor and J. Malda, "Additive Manufacturing of Biomaterials, Tissues, and Organs," *Annals of biomedical engineering*, vol. 45, pp. 1–11, 2017.
4. H. e. a. Wang, "Global, regional, and national life expectancy, all-cause mortality, and cause-specific mortality for 249 causes of death, 1980–2015: a systematic analysis for the Global Burden of Disease Study 2015," *Lancet*, vol. 388, pp. 1459–1544, 2016.
5. P. B. Sick, et al., "Initial Worldwide Experience With the WATCHMAN Left Atrial Appendage System for Stroke Prevention in Atrial Fibrillation," *Journal of American college of cardiology*, vol. 49, pp. 1490–1495, 2007.
6. A. Persona, et al., "Remote control and maintenance outsourcing networks and its applications in supply chain management," *Journal of operations management*, vol. 25, pp. 1275–1291, 2007.
7. J. Savolainen and M. Collan, "Additive manufacturing technology and business model change – a review of literature," *Additive manufacturing*, Vol. 32, Art. 101070, 2020.
8. M. Perini, et al., "Additive manufacturing for repairing: From damage identification and modeling to DLD," *Rapid Prototyping Journal*, 2020 (ahead of print).
9. C. Weller, et al., "Economic implications of 3D printing: Market structure models in light of additive manufacturing revisited," *International Journal of Production Economics*, vol. 164, pp. 43–56, 2015.
10. R. K. Mobley, *An introduction to predictive maintenance*, 2nd ed. Oxford, UK: Butterworth-Heinemann, 2002.
11. D. Thomas, "Costs, benefits, and adoption of additive manufacturing: a supply chain perspective," *International journal of advanced manufacturing technology*, vol. 85, pp. 1857–1876, 2016.

Quantifying the Economic Feasibility of Additive Manufacturing: Simulating Production Lifetimes in the Context of Spare Parts Production

Jyrki Savolainen and Mikael Collan

1 Introduction

As discussed already in the preceding chapters, additive manufacturing (AM) refers to the transferring of digital blueprints into a tangible objects by 3D-printing [1, 2]. The economic feasibility of additive manufacturing has been, and still is a question of discussion—however, it has become clear that there are a number of specific instances, where additive manufacturing is able to create benefits that overweigh the involved costs. Specifically in cases, where the to-be-manufactured items, typically parts, require difficult to construct geometries, or that would benefit from having cavities within the geometry, additive manufacturing has already proven to carry considerable benefits. Discussion in the previous academic

J. Savolainen (✉) • M. Collan
LUT University, Lappeenranta, Finland
e-mail: jyrki.savolainen@lut.fi; mikael.collan@lut.fi

M. Collan, K.-E. Michelsen (eds.), *Technical, Economic and Societal Effects of Manufacturing 4.0*,
https://doi.org/10.1007/978-3-030-46103-4_8

literature has not been very precise about quantifying the economics of additive manufacturing and mostly the discussion has been done on a project-based level, where feasibility of additive manufacturing has been looked at from the point of view of the said projects [3–5]. Industrial contexts that have been studied in this vein include hearing aid production [6], chocolate [7], and military aviation [8].

All in all, the situation is quite unsatisfactory from the point of view of better understanding the types of things that drive the economic feasibility of additive manufacturing in general and there seems to be as such a rather clear need for straightforward quantitative analyses that would illustrate what kind of potential and/or expectations one can make with regards to the economics of additive manufacturing methods, perhaps not alone, but as parts of manufacturing systems. Discussion of additive manufacturing economics has mostly happened within the context of talking about business models around additive manufacturing and more broadly manufacturing 4.0—this is also reflected above in the chapter of this book that concentrates on the business models.

In this chapter we look at the quantification of the feasibility and the economics of additive manufacturing from the point of view of spare parts manufacturing that is we use it as the underlying context. The production of spare parts for technically high-end industries such as the automotive and the aerospace industry [8–11] is something that has already been found economically viable, therefore it constitutes a good ground for quantitative illustrations. Spare parts manufacturing is interesting also from the point of view that the spare parts business has some distinct characteristics, which make it fit especially well with additive manufacturing: there is a need for reducing lead times, for minimizing the supplier's inventory [12], and there is also the issue of extending the time original equipment manufacturers (OEM) are able to offer spare parts support [13, 14]. Typically the demand of spare parts is variable on the very short time frame and the demand trend also changes with the lifecycle of the items for which the parts designed, this is why manufacturers try to rigorously minimize the spare part stock at hand, while they must be able to deliver parts quickly on-demand. The speed requirement is accentuated in situations where the parts-availability is contracted until years ahead.

Traditionally, spare parts inventory-issues are resolved by aggregating demand and serving customers from typically country-specific stocks. As spare parts, for example, for production equipment may be tailor-made for customers and OEM-stock of these parts must be constantly held, this

Table 1 Robust evaluation of economic and manufacturing issues regarding additive and traditional manufacturing (based on previous literature)

Product/Part	*Additive manufacturing*	*Traditional manufacturing*
Market segment (type)	Niche	Mass
Market potential	Low	High
Market volumes	Medium to high	Low to high
Market demand pattern	Stable/uncertain/unitary	Stable/uncertain
Product customization need	High	Low
Product's value added (to customer)	High	Low
Product value versus transport cost	Low	High
Product size	Small	Large
Product's geometric complexity	High	Low
Product's structural integrity requirements	High	Maximum
Manufacturing automation (today)	Low	High
Manufacturing phases	≥ 2	1
Raw material needed to create a product, n	1	≥ 2
Raw material's machinability	Low	High
Raw material cost	High	Low
Raw material origin	Synthetic	Organic
Raw material's natural form	Powder	Solid
Raw material's ratio of material removal (solid only)	High	Low

may mean that the total inventory holding costs of spare parts may be considerably high. Another issue is that when centralized inventories are used, the cost of logistics that is, transportation of spare parts, may be very high, sometimes even higher than the cost of a single part. Table 1 provides a general summary of issues related to the economics and manufacturing of spare parts and products that have been considered important from the point of view of manufacturing economics in the previous literature and for each of which a robust (linguistic) evaluation has been given (also based on previous literature) both for additive and for traditional manufacturing.

The list of issues and the "evaluations", presented in Table 1, has been collected and summarized from [13, 15–25] and makes for a rather comprehensive summary of the things that affect the economies of additive manufacturing in the context of spare parts manufacturing, however the list is most likely relevant also beyond this context. To explore the combination of item qualities, which make them economically viable for AM-technology, we derive and use a simulation model that numerically—instead of qualitatively (see Table 1)—deals with the critically important variables. The type of analysis is of *comparative* type as, parallel to AM-scenario, also the cost of conventional manufacturing is calculated. Assuming an item's quality to be independent on the way that how it is made, it is highly unlikely that an endogenous transition to this new technology would realize without any of the underlying economic drivers, which we now quantify in this research.

With the results from the model we take a stab at the discussion about the "bigger picture" of the economics of additive manufacturing in the context of spare parts manufacturing. The picture we paint is formed taking into consideration the uncertainty and the limited information surrounding the studied six spare part lifetime strategies. The number of previous studies, where simulation analysis is used in this context (economics of additive manufacturing) is small. An exception to the rule is the closely connected simulation-based research by [19] provides insight into postponement strategies in the supply chain in connection with additive manufacturing.

To the best of our knowledge, there is no previous research that would explore the feasibility—and the economic space of additive manufacturing in a similar way that is done here—this makes this research novel in this respect. Here we limit our interest on the economic aspects of using additive manufacturing within the "production lifetime" of spare parts products (production)—more specifically we concentrate on the cost of production aspect. We do not take positions with regards to the different technical aspects and refer the reader interested in technical issues to see [4, 20, 26–31]. We omit discussions about business models surrounding and based on additive manufacturing as they are already discussed in detail elsewhere in this book.

The rest of this paper is organized as follows: the following chapter provides a high-level problem description together with the technical details of the simulation model. then the numerical simulation results from the simulation are presented with factor analysis concerning individual variables. These numbers are supported by a detailed discussion on the attained insights within the limits of the simulation. The paper is closed by drawing some conclusions and discussing the results.

2 Identified Additive Manufacturing Strategies, the Model Used, and the Simulation Setup

The managerial decision on what manufacturing technology is chosen for a new manufacturing "project" is not a fixed one in the sense that the answer is always "conventional" or always "additive". The choice depends on the situation—in a case, where there is no existing demand-base calls for flexibility that is, the ability to be able to start small ("niche market") and then expand production later, in case the demand grows. In a case, where the economics are clear from the beginning, with established markets and stable and strong demand, the easiest and perhaps the safest way to go is typically to choose the use of conventional manufacturing methods. In cases where the demand is low and volatile in the starting phase, then grows and stabilizes, and perhaps towards the end of the life-cycle of the product deteriorates and becomes volatile again, one may consider first using additive manufacturing, then switching to conventional and back again at some point. This means that hybrid strategies are possible, and may be the smartest choice. In this vein, for the purposes of this research we identify six possible production strategies discussed in the previous literature on additive manufacturing:

Strategy 1—"Full TDM", where traditional manufacturing methods are used for the whole manufacturing lifecycle. This is the typical case for products with existing and stable demand and the case, when additive manufacturing possibilities do not exist.

Strategy 2—"Full AM", where the production during the whole manufacturing lifecycle is done by using additive manufacturing technologies.

Strategy 3—"End of life AM", where production is started with traditional manufacturing methods and towards the end of life of the product, when demand typically decreases until it disappears, production is changed to additive manufacturing.

Strategy 4—"Bridge", where the production is started with additive manufacturing methods and if (when) it picks up in a way that justifies using traditional (mass) production technologies and the connected investments they are adopted.

Strategy 5—"Bridge + end of life AM", where production is started with additive manufacturing, then moved to traditional manufacturing and again, towards the end of life changed to additive manufacturing.

Strategy 6—"Other", include more than two production mode switching decisions and therefore does not fall into any of the aforementioned categories.

Now, the goal here is not to test each one of the above strategies separately, by creating a model for them, but to use a numerical simulation to randomly create a large number of "production lifecycles" that exhibit combinations of using AM and TDM during the lifecycle and to see which strategies are manifested in the simulated lifecycles, under what kinds of circumstances, and how often. Practically put, the simulation process to create the production lifecycles is run in the following way for a single lifecycle (scenario):

1. Random values are drawn for the product characteristics from the given variable value ranges (that represent the uncertainty)
2. Month-by-month production-cost arrays for both the additive manufacturing, c_{AM}, and the conventional manufacturing, c_{CM}, of the product are calculated. Cumulative values are calculated by using a fixed demand pattern for 300 months (discussed in detail below). For a more analytical description see Appendix A
3. For each month, we choose the smaller of the two simulated costs $[c_{AM}, c_{CM}]$ to arrive at c_{OPT} that represents a "theoretically optimal production mode" for each month. By adding the monthly c_{OPT} values the cumulative optimal lifetime cost is calculated.
4. The number of switching points is calculated; when $c_{AM}(t) < c_{CM}(t-1)$, and $c_{AM}(t) > c_{CM}(t + 1)$, or vice versa, and the timing of switching with respect to the product lifecycle is observed (see Appendix B for further details)

Put simply, we let the simulation model estimate the "optimal" use of AM-technology by using a simplistic rule of "switch if the alternative is cheaper" and explore the results to see whether the production mode switches coincide with any of the above-listed strategies.

The "world" that underlies the simulation and to which the switching rule is subject to consists of a single (fixed) demand-scenario. This is also (and obviously) a simplification of reality, because in reality there may be a virtually infinite number of possible demand scenarios—but for the sake of illustration, we limit the realism of the simulation and use only one demand scenario. This scenario is based on the idea that the underlying product is a newly launched product, the demand of which has first a fast

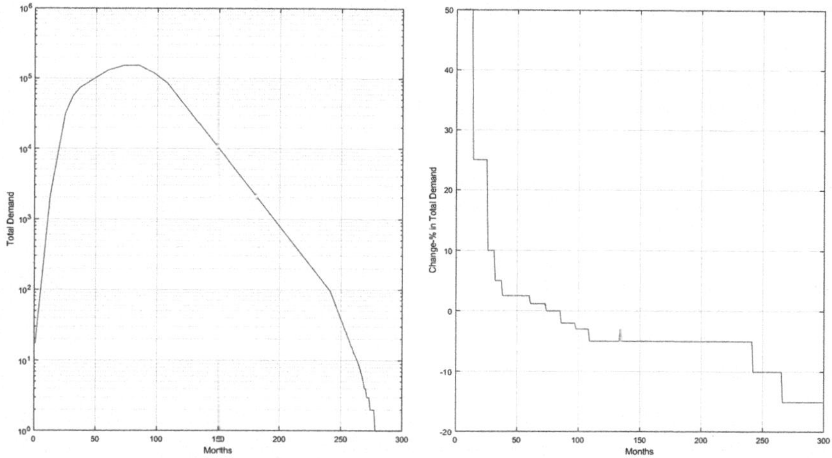

Fig. 1 The demand pattern used, visualized as a function of time. *Left:* Example of total demand of spare parts plotted on a logarithmic y-scale. *Right:* Change-% in demand

positive evolution and that the demand, after having peaked, declines less rapidly. The demand pattern used here is "smooth" and does not include any short-term variance that would in reality be typical of the demand for a product. The demand pattern is visible in Fig. 1.

The rest of "the world" surrounding the product life-time is made of eight variables, see Table 2, the values of which, together with the demand curve can be used to determine whether the production cost is lower with additive manufacturing or by way of traditional manufacturing. The variables represent issues that have in the literature been identified as important from the point of view of adoption of additive manufacturing—the selection has been made by the authors. To generalize the cost of additive manufacturing is defined as a multiplier of traditional manufacturing unit cost.

In the simulation each scenario (each production situation tested) is randomly generated, by drawing a random value for each one of the input variables. The initial variable value distributions are uniform that is, all values are equally likely for all parameters. After the initial values are drawn the "production" continues month-by-month following the demand curve. What we effectively do is that we run a Monte Carlo simulation

Table 2 Input variables and their limits, uniform distributions are used

Variable	*Description*	*Unit*	*Min*	*Max*
i1	Installed base of existing equipment at $t = 0$	pcs	10	1000
i2	Average part lifetime	months	6	36
i3	Safety Stock requirement	– [ratio]	0.01	1
i4	Relative direct cost of part using AM vs. TDM	– [ratio]	1	10
i5	Cost of part using TDM	– [$/€/£, etc.]	10	1000
i6	Minimum lot size of TDM	pcs	10	5000
i7	Relative cost of storing in TDM	– [ratio]	0.01	0.1
i8	Riskless interest rate	%	0.01	0.1

(MCS)—the number of rounds or "productions" used is 500,000. The resulting input-output pairs are then studied by way of factor analysis (FA) to identify the robust relative importance of each variable for the outcome. As such the results are quite robust and general.

The Simulation Model Used

As discussed above, the main idea of using a simulation model is to perform a robust comparison of the relative costs of additive manufacturing versus conventional manufacturing and to see whether it is AM or TDM that would be used in the production. In essence, the model calculates the monthly production cost for the both production methods, based on selected input variable values and a fixed demand curve. At any point of time during the simulated production life-span, the preferred production mode is explicitly given as:

$$\mathbf{IF}\left[\text{current AM cost}\right] < \left[\text{current TDM cost}\right] \\ : \mathbf{THEN}\ \text{use AM}\ \mathbf{ELSE}\ \text{use TDM}. \quad (1)$$

We assume that the demand is fulfilled instantly when AM is used, whereas in the case of TDM items must be ordered beforehand—a *safety stock requirement* (*i3*) is used to buffer the demand in TDM, which increases the total number of spare parts needed (for details see Appendix A). Based on the *relative cost of storing* (i7) a total yearly inventory cost is calculated, which is then evenly divided to months and used in the monthly

Table 3 Outputs from the simulation model

Variable	*Description*	*Unit*
o1	Initial mode of production, where 0 = TDM and 1 = AM	0/1
o2	Sum of AM months	Months
o3	Total number of switching points	*N*
o4	Time of first production mode switch	– (time)
o5	Time of the last production mode switch	– (time)
o6	Final mode of production	0/1
o7	Product volume	*N*
o8	TDM total cost	– [$/€/£, etc.]
o9	AM total cost	– [$/€/£, etc.]
o10	Absolute cost difference, TDM vs. AM	– [$/€/£, etc.]
o11	Relative cost difference, TDM vs. AM	%

EoL end of life, *TDM* traditional manufacturing, *BrM* bridge manufacturing

cost calculation. ***Riskless interest rate*** (i8) describes the opportunity cost of having the money invested into product stock. As additional simplification, we assume the production of spare parts to be fully contracted and requiring no capital investment on the production equipment and that there are no restrictions in the availability of the contracted capacity. No obsolescence of stored products is assumed in modeling the TDM cost.

The list of outputs is shown in Table 3. Based on Eq. (1) we count the *total number of AM months* (*o2*) for each simulated production lifetime. The information on the number of production mode switches during the simulation is stored in output *o3*, outputs *o7-o11* are calculated from the simulation results. The focal output of this exploratory research is the costs of AM compared to the costs of TDM over the production lifetime and most importantly whether AM or TDM is used—the technology that is used determines the additive manufacturing profile of the production lifecycle that is compared to the six identified strategies. Also the (expected) cumulative costs for both TDM (*o8*) and AM (*o9*) are calculated for the simulation. To determine the "preferred mode of production" we have:

$$\textbf{IF}\left[\text{cumulative AM cost}\right] < \left[\text{cumulative TDM cost}\right]: \textbf{THEN}\text{ select} \left[\text{Preferred mode of production}\right] = \text{AM}\,\textbf{ELSE}\,\text{TDM} \quad (2)$$

The first and the last switching point times (*o4 and o5*), together with production mode information (*o1 and o6*) are combined into an insight of switching flexibility. These outputs indicate when, if at all, AM-technology would most likely be used during the production lifetime. For simplicity, we assume that the production mode (AM/TDM) can be changed. Costs and delays and that the inventory carrying costs are not inherited, once the production changes from TDM to AM. We observe that this is a non-realistic simplification and do not claim that the model gives a fully realistic picture of how switching could happen in the real world, where hysteresis is typically present in switching decisions [32]. We point out that the definitions used for the six reference-strategies are provided in Appendix B.

3 Simulated Results and Analysis

The simulated results for the 500,000 production lifecycles were automatically processed according to the rules laid out in Appendix B and matched to the six reference additive manufacturing strategies. The results are visible in Table 4.

Based on the obtained results we are able to summarize some exploratory insights. First, there seems to be potential to extend the lifetime of products with AM-based spare part support—even in situations, where the cost of AM is significantly higher than the cost of TDM (up to 5.76∗x)—this is in line with Strategy 3, "End of life AM". Second, Strategy 4 "Bridge"-manufacturing applications are found seem to be viable, when a small installed base of equipment exists. We further note that investing in bridge manufacturing effectively opens the option for end-of-life additive manufacturing. The combination of bridge—and end-of-life manufacturing that is Strategy 5, is visibly a more prominent manufacturing strategy than bridge-manufacturing alone. Third, high-volume products with reasonably small LOT-sizes that do not have manufacturing or inventory cost disadvantages, seem to stay mass-manufactured that is, Strategy 1 is strong for products of this type.

Interestingly it seems that also high-volume products can benefit from end-of-life AM that is, Strategy 3 is strong also there. Overall Strategy 3 and end-of-life production is the leading strategy in terms of where additive manufacturing seems feasible. Strategy 2, using additive manufacturing for the whole production life cycle, seems to be marginal in terms of how frequently it is feasible in the context of this study—only 0.12% of

Table 4 Results of 500,000 rounds of simulations divided strategy by strategy and mean values for the input variables and for the outputs

Mean values	EoL	*TDM*	BrM + EoL	*Others*	BrM	*Full AM*
Sum of AM months	26.82	0.00	59.84	16.00	2.00	301.00
Total Number of Switches	1.00	0.00	3.26	2.54	1.00	0.00
Time of First Switch	275.18	NaN	4.22	265.61	3.00	0.00
Time of Last Switch	275.18	NaN	248.33	286.76	3.00	0.00
Final mode (AM = 1; Trad = 0)	1.00	0.00	1.00	0.33	0.00	1.00
Time of 1st Switch (cumul.)	NaN	NaN	5.73	5.09	3.94	NaN
Volume, *n*	32.83	32.21	12.21	26.76	3.50	29.74
TDM-cost, [a]1000 unit of money	16,742.15	16,422.77	6206.34	13,668.80	1749.90	15,371.64
AM-cost, [a]1000 unit of money	83,355.06	107,873.65	8904.56	60,140.40	4724.97	15,219.94
Diff AM vs. TDM, [a]1000 unit of money	66,612.91	91,450.88	2698.22	46,471.60	2975.06	−151.70
Diff-% AM vs. TDM	4.14	5.59	1.36	3.51	3.69	−0.01

EoL end of life, *TDM* traditional manufacturing, *BrM* bridge manufacturing

Bolded numbers indicate the most significant input variables based on the factor analysis

outcomes reflected this strategy. In these cases the cost of AM very close to the cost of TDM.

To further study the flexibility to switch the production mode, some key results are illustrated in Fig. 2. A great majority of the simulated production lifecycles have between zero and two production mode changes (96.87%)—we feel that this shows that the simulation model is in this respect quite realistic. The remaining Strategy 6 that is, "*other*" scenarios represent only 3.13% of the outcomes and seem to have an unrealistic average of 18 production mode changes, where outliers have even many tens of changes—as discussed above, this is a feature of the simulation structure and can be explained by the fact that issues such as hysteresis are not considered. The first transition from AM to TDM (or vice versa)

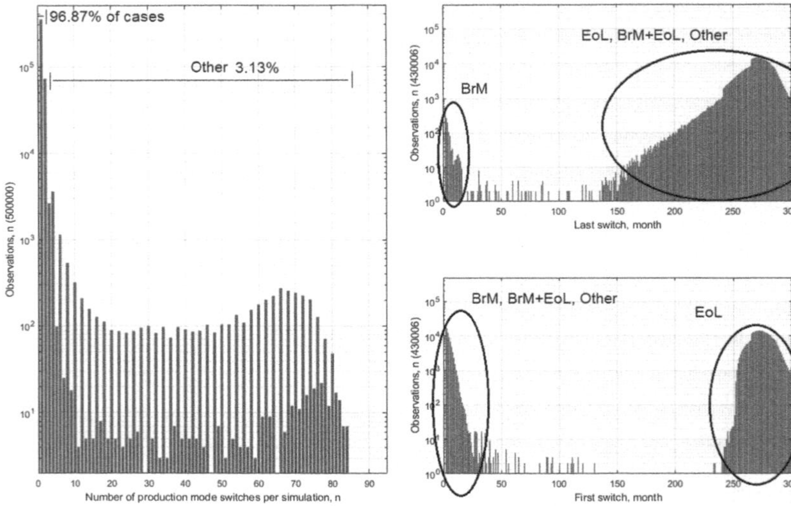

Fig. 2 Histogram representation of production-mode changes from the 500,000 simulation rounds. Observe that the y-axes are logarithmic. *Left:* # of mode changes per simulation. *Upper right:* Timing of the first change. *Lower right:* Time of the last change

occurs either in the start, or in the end of product lifetime. The timing of the last change exhibits an interesting pattern that suggests that the timing of the final production mode change in Strategy 3 (End-of-life AM) depends on product-specific characteristics and the switch time ranges from ~150 to 300 months and averaging for Strategy 3 at 274 months (~8.5% of life left) and for Strategy 5 (Bridge + end-of-life) at 247 months (~17.5% life left). The bridge-manufacturing period lasts on average only four months (1.5% of life). These simulated change-times are dependent on the inputs and especially on the demand curve, so they cannot be generalized and are quoted here for illustration only (Table 5).

Factor analysis of the results reveals that some included input variables have only a small effect on the economic feasibility of AM (parts durability, safety stock ratio, TDM unit cost, and inventory costs). On the other hand the relationship between the cost of AM and the cost of TDM and the size of the installed machine base (production volume) seem to be important issues from the point of view of AM feasibility. These are, however, only

Table 5 Factor analysis of the input variables

	Strategy 1 Full TDM	*Strategy 2 Full AM*	*Strategy 3 EoL*	*Strategy 4 BrM*	*Strategy 5 BrM + EoL*	*Strategy 6 Other*
i1 Installed base	0.043	−0.044	−0.086	**0.997**	**0.997**	−0.038
i2 Durability, months	−0.036	−0.015	0.076	0.069	0.156	0.021
i3 Safety stock ratio	−0.01	*0.351*	0.001	−0.206	−0.013	−0.014
i4 AM cost multiplier	0.095	**0.997**	**0.997**	*−0.51*	*−0.481*	**0.916**
i5 TDM Unit Cost	−0.001	0.044	0.001	0.048	0.008	0
i6 TDM LOT-size	**0.997**	0.05	0.134	0.177	0.143	−0.273
i7 TDM Storage cost ratio	−0.223	0.239	0.136	0.232	0.103	−0.108
i8 Riskless interest Rate	0.001	0.285	0.002	−0.034	−0.005	0.009

EoL end of life, *TDM* traditional manufacturing, *BrM* bridge manufacturing

Value-range is [−1, −1], where values close to 1 and −1 indicate a strong relationship (direct or "inverse") with the Strategy. Inverse means here that large negative values mean that the small variable values link to the specific Strategy. Strong relationships bolded, important relationships in italics

exploratory and illustrative results that are the results of the assumptions used in the simulation model.

4 Summary, Conclusions, and Discussion

This chapter has discussed the economic viability and feasibility of additive manufacturing as a part of the production lifecycle of spare parts production. A simulation model was used to randomly create a large number of different possible production lifecycles that were, with the help of a stylized demand curve, analyzed for, when under the assumed circumstances production would be made with additive manufacturing and alternatively with traditional manufacturing methods. The resulting lifecycle patterns of AM and TDM use or manufacturing strategies were compared to six, from the literature identified strategies, and the relative frequency of the

Table 6 Main findings for the six production strategies with frequency in the simulations

Production strategy	*Strategy details, relative frequency in simulations*
Strategy 1—"Full TDM"	The main strategy for high-volume, cheap-to-store products with reasonably small LOT-sizes; 14% of the cases
Strategy 2—"Full AM"	Strategy seems to be feasible only, when AM manufacturing costs are ~equal to the costs of TDM; 0.12% of the cases
Strategy 3—"End of life AM"	Strategy for parts with relatively large LOT-sizes that are too expensive for large scale AM production. With declining demand, the costs of holding inventories outweigh the costs of AM-production; 68.5% of the cases
Strategy 4—"Bridge"	Strategy for niche-volume parts with very high durability and large LOT-sizes. AM is feasible for a short period of time after the product launch; 0.23% of the cases
Strategy 5—"Bridge + End-of-Life AM"	Strategy for small volume, highly durable parts, which would require large LOT-sizes in mass production, but are relatively cheap to manufacture with AM; 14% of the cases
Strategy 6—"Other"	Strategies for small to medium volume parts that are both cheap to produce with AM and have a small LOT-size in TDM, "the rest of the strategies"; 3.13% of the cases

six strategies was determined. Table 6 lists the main findings for each of the six strategies.

The findings presented above are in line with what can be found from the literature and quite strongly reinforce the notion that end-of-life use of additive manufacturing is an economically viable and feasible strategy to adopt the use of additive manufacturing in the production lifecycle of spare parts. Factor analysis was used to find the most important variables from the point of view of each one of the strategies—the main finding was that the relationship between the cost of AM and the cost of TDM and the size of the installed machine base (production volume) seem to be important issues from the point of view of AM feasibility.

This research is exploratory and has numerous limitations—the results are guided by the initial selection of variables, their value ranges, and the fixed demand pattern used. The simulation was performed purely with a simple cost point-of-view in mind and without taking into account other factors. Issues such as design benefits and other added value from AM, market dynamics, hysteresis, and many other possibly important issues were left outside the scope of this research. This being said, the research presented here is among the first attempts to quantify the feasibility of additive manufacturing in production lifecycles and as such is a contribution to what we know about the economics of additive manufacturing.

Appendix A

DEMAND MODELING/Actual and safety stock demand for spare parts at time t:

[Demand actual, *t*] = [Installed base, *n*] × [Part durability, months]
[Demand stock, *t*] = [Demand actual, *t*] × [Safety stock ratio]

- [Demand TOTAL, *t*] = [Demand actual, *t*] + [Demand stock, *t*]

COST MODELING/Cost of conventional manufacturing:

Total cost of conventional manufacturing, C_{CM}, is the sum of production cost, stock holding cost and the opportunity cost of holding the stock:

$C_{CM,p}$ = [Production Lot-size, *n*] × [Demand TOTAL, *t*] × [Production cost CM, EUR/pc]

$C_{CM,s}$ = [Stock size, pcs] × [Stock holding cost, EUR/pc] + [Opportunity cost, EUR], where

[Opportunity cost, EUR] = [Production cost CM, EUR/pc] × [Stock holding cost, EUR/pc] × [Riskless Interest Rate]

- C_{CM} = ([Production Lot-size, *n*] × [Demand TOTAL, *t*] × [Production cost CM, EUR/pc]) + ([Stock size, pcs] × [Stock holding cost, EUR/pc]) + ([Production cost CM, EUR/pc] × [Stock holding cost, EUR/pc] × [Riskless Interest Rate])

COST MODELING/Cost of additional manufacturing:

Total cost of additive manufacturing does not include the costs associated with production lot sizes and stocks:

C_{AM} = [Demand TOTAL, *t*] × [Production cost, AM]

Appendix B

Analytical definitions of the six production strategies

Strategy 1/2—"Full AM/TDM":

$$\begin{aligned}&\textbf{IF}\left[\text{Number of swithces}\right]=0\\&\textbf{AND}\left[\text{Final production mode}\right]=\left[\text{AM}\right]\left(\text{TDM}\right)\\&\textbf{AND}\left[\text{Initial production mode}\right]=\left[\text{AM}\right]\left(\text{TDM}\right)\end{aligned} \tag{8.3, 8.4}$$

Strategy 3—"End-of-Life AM":

$$\begin{aligned}&\textbf{IF}\left[\text{Number of switches}\right]=1\\&\textbf{AND}\left[\text{Final production mode}\right]=\left[\text{AM}\right]\\&\textbf{AND}\left[\text{Initial production mode}\right]=\left[\text{TDM}\right]\end{aligned} \tag{8.5}$$

Strategy 4—"Bridge Manufacturing":

$$\begin{aligned}&\textbf{IF}\left[\text{Number of switches}\right]=1\\&\textbf{AND}\left[\text{Final production mode}\right]=\left[\text{TDM}\right]\\&\textbf{AND}\left[\text{Initial production mode}\right]=\left[\text{AM}\right]\end{aligned} \tag{8.6}$$

Strategy 5—"Bridge and End-of-Life manufacturing":

$$\begin{aligned}&\textbf{IF}\left[\text{Number of switches}\right]=2\\&\textbf{AND}\left[\text{Initial production mode}\right]=\left[\text{AM}\right]\\&\textbf{AND}\left[\text{Final production mode}\right]=\left[\text{AM}\right]\end{aligned} \tag{8.7}$$

Strategy 6—"The remaining production strategies—others", are derived in two parts: (a) scenarios, where the number of production mode changes is greater than two; and (b) cases that have two production mode changes, but are not Bridge + End-of-life (AM start and end):

$$\textbf{IF}\left[\text{Number of switches}\right]>2 \tag{8.7a}$$

and

$$\textbf{IF}\left[\text{Number of switches}\right]=2$$

$$\textbf{AND}\left[\text{Initial production mode}\right]=\left[\text{TDM}\right]$$
$$\textbf{AND}\left[\text{Final production mode}\right]=\left[\text{TDM}\right] \quad (8.7b)$$

References

1. T. Rayna and L. Striukova, "From rapid prototyping to home fabrication: How 3D printing is changing business model innovation," *Technol. Forecast. Soc. Change*, vol. 102, pp. 214–224, 2016.
2. J. Holmström, M. Holweg, S. H. Khajavi, and J. Partanen, "The direct digital manufacturing (r)evolution: definition of a research agenda," *Oper. Manag. Res.*, vol. 9, no. 1–2, pp. 1–10, 2016.
3. M. Attaran, "The rise of 3-D printing: The advantages of additive manufacturing over traditional manufacturing," *Bus. Horiz.*, vol. 60, no. 5, pp. 677–688, 2017.
4. T. D. Ngo, A. Kashani, G. Imbalzano, K. T. Q. Nguyen, and D. Hui, "Additive manufacturing (3D printing): A review of materials, methods, applications and challenges," *Compos. Part B Eng.*, vol. 143, pp. 172–196, 2018.
5. V. Petrovic, J. Vicente Haro Gonzalez, O. Jordá Ferrando, J. Delgado Gordillo, J. Ramon Blasco Puchades, and L. Portoles Grinan, "Additive layered manufacturing: Sectors of industrial application shown through case studies," *Int. J. Prod. Res.*, vol. 49, no. 4, pp. 1061–1079, 2011.
6. C. G. Sandström, "The non-disruptive emergence of an ecosystem for 3D Printing – Insights from the hearing aid industry's transition 1989-2008," *Technol. Forecast. Soc. Change*, vol. 102, pp. 160–168, 2016.
7. F. Jia, X. Wang, N. Mustafee, and L. Hao, "Investigating the feasibility of supply chain-centric business models in 3D chocolate printing: A simulation study," *Technol. Forecast. Soc. Change*, vol. 102, pp. 202–213, 2016.
8. S. H. Khajavi, J. Partanen, and J. Holmström, "Additive manufacturing in the spare parts supply chain," *Comput. Ind.*, vol. 65, pp. 50–63, 2014.
9. J. Holmström, J. Partanen, J. Tuomi, and M. Walter, "Rapid manufacturing in the spare parts supply chain: Alternative approaches to capacity deployment," *J. Manuf. Technol. Manag.*, vol. 21, no. 6, pp. 687–697, 2010.
10. L. F. C. S. Durão, A. Christ, R. Anderl, K. Schützer, and E. Zancul, "Distributed Manufacturing of Spare Parts Based on Additive Manufacturing: Use Cases and Technical Aspects," *Procedia CIRP*, vol. 57, no. January 2017, pp. 704–709, 2016.
11. M. Savastano, C. Amendola, F. D'Ascenzo, and E. Massaroni, "3-D Printing in the Spare Parts Supply Chain: An Explorative Study in the Automotive Industry," in *Lecture Notes in Information Systems and Organisation*, vol. 18, 2016.

12. S. Ford and M. Despeisse, "Additive manufacturing and sustainability: an exploratory study of the advantages and challenges," *J. Clean. Prod.*, vol. 137, pp. 1573–1587, 2016.
13. G. Manogharan, R. A. Wysk, and O. L. A. Harrysson, "Additive manufacturing-integrated hybrid manufacturing and subtractive processes: Economic model and analysis," *Int. J. Comput. Integr. Manuf.*, vol. 29, no. 5, pp. 473–488, 2016.
14. M. Gebler, A. J. M. Schoot Uiterkamp, and C. Visser, "A global sustainability perspective on 3D printing technologies," *Energy Policy*, vol. 74, no. C, pp. 158–167, 2014.
15. C. Weller, R. Kleer, and F. T. Piller, "Economic implications of 3D printing: Market structure models in light of additive manufacturing revisited," *Int. J. Prod. Econ.*, vol. 164, pp. 43–56, 2015.
16. M. Matsumoto, S. Yang, K. Martinsen, and Y. Kainuma, "Trends and research challenges in remanufacturing," *Int. J. Precis. Eng. Manuf. Technol.*, vol. 3, no. 1, pp. 129–142, 2016.
17. B. Berman, "3-D printing: The new industrial revolution," *Bus. Horiz.*, vol. 55, no. 2, pp. 155–162, 2012.
18. J. P. J. de Jong and E. de Bruijn, "Innovation Lessons From 3-D Printing," *MIT Sloan Manag. Rev.*, vol. 54, no. 2, pp. 43–52, 2013.
19. J. Minguella-Canela et al., "Comparison of production strategies and degree of postponement when incorporating additive manufacturing to product supply chains," *Procedia Manuf.*, vol. 13, pp. 754–761, 2017.
20. M. Bogers, R. Hadar, and A. Bilberg, "Additive manufacturing for consumer-centric business models: Implications for supply chains in consumer goods manufacturing," *Technol. Forecast. Soc. Change*, vol. 102, pp. 225–239, 2016.
21. M. Rehnberg and S. Ponte, "From smiling to smirking? 3D printing, upgrading and the restructuring of global value chains," *Glob. Networks*, vol. 18, no. 1, pp. 57–80, 2018.
22. A. Laplume, B. Petersen, and J. M. Pearce, "Global value chains from a 3D printing perspective," *J. Int. Bus. Stud.*, vol. 47, no. 5, pp. 595–609, 2016.
23. M. J. Ryan et al., "3D printing the future : scenarios for supply chains reviewed," *Int. J. Phys. Distrib. Logist. Manag.*, vol. 47, no. 10, pp. 992–1014, 2017.
24. B. P. Conner et al., "Making sense of 3-D printing: Creating a map of additive manufacturing products and services," *Addit. Manuf.*, vol. 1–4, pp. 64–76, Oct. 2014.
25. Wohlers, "Additive Manufacturing Technology Roadmap for Australia," 2011.
26. T. Caffrey and T. Wohlers, "An Additive Manufacturing Update," *Appl. Des.*, no. May, pp. 27–29, 2016.
27. K. V. Wong and A. Hernandez, "A Review of Additive Manufacturing," *ISRN Mech. Eng.*, pp. 1–10, 2012.

28. I. Gibson, D. W. D. W. Rosen, and B. Stucker, *Additive Manufacturing Technologies: Rapid Prototyping to Direct Digital Manufacturing*. Springer, 2009.
29. T. Wohlers and T. Gornet, "History of additive manufacturing," *Wohlers Rep. 2014*, p. 34, 2014.
30. W. Gao et al., "The status, challenges, and future of additive manufacturing in engineering," *Comput. Des.*, vol. 69, pp. 65–89, Dec. 2015.
31. H.-J. Steenhuis and L. Pretorius, "Consumer additive manufacturing or 3D printing adoption: an exploratory study," *J. Manuf. Technol. Manag.*, vol. 27, no. 7, pp. 990–1012, 2016.
32. A. K. Dixit, "Investment and Hysteresis," *J. Econ. Perspect.*, vol. 6, no. 1, pp. 107–132, 1992.

Industry 4.0. Transformation Challenge in Light of Dynamic Capabilities

Kalevi Kyläheiko and Päivi Maijanen

1 Introduction

The manufacturing industry is changing drastically due to the changes brought about by the driving forces of the so-called Industry 4.0 Transformation, such as digitalization, modularization, additive manufacturing (3D printing), robotics, artificial intelligence, mass customization, global keen rivalry, etc. Perales et al. [1] characterize Industry 4.0. by means of virtualization, interoperability, automation, flexibility, real-time availability, service orientation, and energy efficiency, whereas Zezulka et al. [2] emphasize (1) digitization and integration of

K. Kyläheiko (✉)
University of Bremen, Bremen, Germany

LUT University, Lappeenranta, Finland
e-mail: kalevi.kylaheiko@lut.fi

P. Maijanen
LUT University, Lappeenranta, Finland
e-mail: paivi.maijanen-kylaheiko@lut.fi

M. Collan, K.-E. Michelsen (eds.), *Technical, Economic and Societal Effects of Manufacturing 4.0*,
https://doi.org/10.1007/978-3-030-46103-4_9

networks, (2) digitization of product and services, and (3) generating new market (preferably business) models. This means that the firms that want to sustain their competitive advantage (CA) have to be able to rapidly reorganize and transform their resource (especially knowledge) bases and operational manufacturing skills, routines, and capabilities in a cost efficient way. Large multinational companies as well as more local small and medium-sized enterprises in the manufacturing industry have to proactively find a way to re-invent their new business models in a way that combines their strengths in product design and manufacturing together with existing, or to-be-created new capabilities in managing digital ecosystems. Because of strong network externalities (resulting in "the winner takes it all"-earning logic) one can anticipate that only the companies that are the first ones to reinvent, recombine and finally standardize new capability combinations as necessary bottleneck complementary assets will be able to sustain their CA.

Teece [3] uses "*digital convergence*" as an umbrella concept that touches upon the main aspects of the digital revolution. In its core is wireless communication based on digital broadband technologies. It makes it possible to effectively and flexibly control and monitor extensive platforms and digital ecosystems. Digital information is not locally constrained anymore, which means that the firms can flexibly locate their manufacturing activities so that they can best satisfy individual needs of buyers/customers. The sensors, microprocessors, learning algorithms, etc. enable the firms to remote control complex supply chain networks and manufacturing design problems as well as to anticipate of potential problems already before anything fatal takes place (the rapid rise of the so called "Internet of Things, IoT" manifests that). This means that large parts of manufacturing products can be designed where the best high-tech expert knowledge exists and then flexibly produced where the demand is, for instance, by means of additive manufacturing (3D printers) and robots. Because the role of manual unskilled workers becomes less important, manufacturing becomes more foot-loose thus allowing the rise of "born globals" even amongst small and middle-sized enterprises. Because of digitalization of production design, also the border line between products and services partly loses its importance. The manufacturing firms are also service providers—the integration of manufacturing and service is called servitization in [4] and they show empirically that there seems to be a nonlinear U-shaped interaction-effect between digitalization and servitization on financial performance in a sample of 131 manufacturing companies. In addition, the industry boundaries are losing their distinctive nature and, in fact, many of the most promising opportunities can be found on the interfaces between former industry clusters.

Information technologies, banking and finance as well as retailing have been the forerunners in digital convergence. They have managed to put together internet, wireless communication and all kinds of wireless services from music, movies, and cameras to social media (see [3]). The same concept, internet + retailing (Amazon, Alibaba, and eBay), internet + hoteling (AirBnB), internet + taxi (Uber) seems to be working also in other services. They all are also good examples of how modern platforms and digital ecosystems utilize positive network externalities and how the "winner takes it all logic" works.

The manufacturing industry has been much slower at exploiting the huge opportunities of digitalization and wireless communication but it is clear that we will quite soon see also there the rise of digital information based manufacturing ecosystems i.e. the rise of Internet 4.0. However, it is not at all clear who will be the winners and losers within these new ecosystems. The sad histories of Kodak and Polaroid tell the story of how an industry leader can rapidly lose its position when facing the challenge of digitalization, if the management is myopic and unable to respond the new challenges [3, 5, 6]. It may happen that the new leaders will come from outside. For instance, some recent endeavors of Google clearly show that they are eager to take steps to this direction.

Even if we will not go deeper into social issues of Industry 4.0 in this section, it is worth noting one important implication of Industry 4.0.Transformation. The extensive adoption of advanced manufacturing methods will necessarily result in large re-allocations of global labor force working nowadays in manufacturing companies. First, extensive robotics, great flexibility and remote wireless control of digitally designed products/services mean that the role of unskilled labor will be diminishing at the same time, as the role of skilled labor and high-tech experts will be increased. Second, we will also see large geographical relocations, since the importance of labor costs, i.e. wage differentials between countries will not be so important determinants of the location decision as they used to be. For instance, if in an emerging country, let's say India, the labor cost is 50% lower than the labor cost of Germany and the share of labor cost is 40% of the value of the product, then the cost advantage of manufacturing the product in India is 20%. This cost advantage most likely covers all the extra re-location costs (transaction costs included) and results in outsourcing manufacturing activities to India. However, if the launching of robots etc. reduces the share of manufacturing labor costs from 40% to 10%, then the cost advantage is only 5% that hardly covers all extra cost of outsourcing. In this situation, there are no incentives anymore to outsource and we will likely see global supply networks becoming more regional again, more about this, see [7]. This again means that it is most likely that outsourcing is not playing a great role during the era of Industry 4.0. Much more important is the location of high-tech science-based knowledge. This will make advanced manufacturing companies foot-loose and the so-called high wage rate countries will be winners.

Because of the new phenomena and mechanisms that digital convergence creates for the manufacturing firms, it is of great importance to try to understand how they could sustain their CA also in the digital era. In order to give answers to this fundamental question, we will look at the tools that modern strategic management can offer to successfully overcome the transformation challenge. The main challenge that the companies are facing is the challenge of organizational renewal under the circumstances of radical uncertainty. After carefully analyzing the pros and cons of different approaches, we will conclude that the ***dynamic capability*** *view (DCV)* launched by Teece and others [8] is the most suitable approach to analyze the challenge. When dealing with the Industry 4.0 Transformation from the managerial and organizational perspective one can conclude that it is mainly about how to create dynamic capabilities that are able to change path-dependent operational manufacturing capabilities and resource bases in a way that enables a company to sustain its CA [8–11].

This article is organized as follows. First, we will briefly analyze the different approaches of strategic management and look at their general managerial implications. Then we will focus on DCV. Especially its micro foundations will be stressed. In addition, the importance of the Schumpeterian [12] entrepreneurial attitude, i.e., the ability to create "new combinations" as an important precondition to overcome transformation challenges will be discussed. The next section goes further and deeper and utilizes the Teecean sensing-seizing-reconfiguring framework in the context of the digital ecosystems. The main question here is how to profit from innovation in networked ecosystems faced by the firms of the Industry 4.0. We will look at the ways how the firms can create and capture value in these conditions where new kinds of dynamic capabilities and new business models are needed. Finally, some important managerial implications and conclusions concerning the ways to overcome the transformation challenge of Industry 4.0. will be offered.

2 On Different Strategic Management Approaches When Facing the Industry 4.0. Transformation Challenge

All the economics-based strategic management approaches attempt to answer the fundamental question of how to achieve and sustain CA i.e. why some firms are able to outperform others. Basically, there are three explanations for sustainable extra profits (or rents):

(a) *Monopoly-based rents* are based on product or service market imperfections and the main strategic message is to position a firm so that it maximizes its monopoly (bargaining) power at the same time as it minimizes the monopoly efforts of rival companies. Porter [13, 14] brought these ideas into strategy research by means of his famous "*five forces model*".
(b) *Scarcity-based rents* in turn are based on factor or resource market imperfections (instead of Porterian product market imperfections). Following the old ideas of David Ricardo [15], the resource-based view (RBV) posed this issue in the mid 1980s in strategy research [16–18]. Barney [19] summarized the basic managerial message as follows: try to base your competitive advantage on the resources with V(valuable), R(rare), I(inimitable), and N(non-substitutable) attributes. In other words, a firm is able to sustain CA, if it employs resources that create value (meaning that someone is willing to pay for their services) and are scarce and hard to imitate and substitute.
(c) *Entrepreneurial rents* are based on the firm's ability to find Schumpeterian [12] new combinations i.e. to utilize its resource and knowledge bases in a new way that create new earning opportunities. The dynamic capability view (DCV) introduced by Teece and others [8] opened up this evolutionarily inspired way of thinking in modern strategy research. The most important difference when compared to the Porterian or resource-based view is the dynamic nature of this approach.

Next, we will briefly discuss the pros and cons of three different CA explanations in the context of Industry 4.0. Transformation.

(a) The Porterian *five forces model* is based on the microeconomics-based monopoly model and tries to maximize the bargaining power of the firm. Porter [13, 14] introduces three strategies to obtain CA: (1) *cost advantage strategy* based on economies of scale and scope, (2) *differentiation strategy* based on the ability to create brands with inelastic demand, and finally (3) *niche-based strategies* suitable mainly for small and middle-sized enterprises. When analyzing these strategies in the context of the Industry 4.0. Transformation we are tempted to argue that traditional production-related economies of scale are not of great importance any more, since they are more suitable for the industry model where decreasing average total costs could be achieved by large conglomerates. However, also Industry 4.0. offers economies of scale-based advantages mainly for two reasons that in fact explain the rise of platform-based ecosystems during the last 10–15 years. First, digital products/services are often characterized by high first copy cost and then rapidly decreasing marginal costs often approaching zero. This combination creates strictly decreasing average costs and, consequently, a decreasing supply curve. Second, digital goods are also characterized by strong demand-related positive network externalities resulting in the increasing demand curve (up to a certain point). Together these two elements often lead to the "winning takes it all" equilibrium where one company or few oligopolies dominate global markets (think about Google, Facebook, Amazon, Alibaba, Airbnb, etc.). Hence, the main lesson for the companies facing the Industry 4.0. Transformation is to try to simultaneously utilize both the decreasing average costs and (up to a certain point) increasing demand curve.

In addition, the economies of scope are of importance in the era of Industry 4.0, too. If a company is very good at doing something special because of its strong core capabilities, they should try to find other industries (or in fact platforms/ecosystems) in which they can apply them as well. Flexibility, digitalization and globalization of the new industrial world create many new opportunities to exploit this potential. Differentiation strategy can also be utilized during the Industry 4.0. era, since the digitalization/mass customization/servitization all create more opportunities to be global instead of being regional as it used to be in the older manufacturing model. The same holds true for the niche-creation strategy. There are opportunities for agile "born globals" as well.

(b) Next, we will have a look at the interpretations offered by the RBV. As mentioned, it is based on the land rent ideas of Ricardo from the year 1817. Instead of focusing on product market imperfections RBV focuses on factor market imperfections. If a company manages to have resources with VRIN (valuable, rare, inimitable, non-substitutable) attributes, it is able to have at least temporary CA [19]. The stronger the so-called isolation mechanisms based, for instance, on causal ambiguity or tacit knowledge [16] are, the better the company is able to establish sustainable CA. Unfortunately, it seems that during the era of Industry 4.0., the opportunities to base CA solely on tangible VRIN resources seem to be very limited. However, the opportunities to utilize knowledge-based intangible assets as elements of CA are much higher. This advantage often utilizes strong and effective software algorithms in order to create totally new customer-tailored services with strong positive network externalities discussed before. In our view, the main message in the context of the Industry 4.0. Transformation has to be reanalyzed and rewritten but, clearly, the intangible resources with VRIN attributes still remain as important sources for CA.

(c) Finally, entrepreneurial Schumpeterian rents that can be obtained and sustained by means of dynamic capabilities to renew, rethink, create and destroy existing resource and knowledge bases to better respond to the challenges of rapid environmental changes are of great importance when trying to face the challenges of the Industry 4.0. Transformation.

After this introduction to the basic ideas of modern strategic management that show that they all are relevant when trying to utilize strategic options created by the Industry 4.0. Transformation, we will now go further by concentrating mainly on dynamic capabilities as main sources of entrepreneurial Schumpeterian rents. Nevertheless, we have to keep in mind that the strategic elements revealed by the Porterian and RBV have to also be taken into account.

3 Dynamic Capabilities: What Are They All About

When analyzing the importance of DCV it is advisable to start with a broader picture that sheds light on its evolutionary roots. In their influential book, "Evolutionary Theory of Economic Growth" Nelson and Winter [20] launched the idea about the firms that consist of routines and more collective bundles of routines, called capabilities. They are stable learned patters that enable a company to be successful. Because of bounded rationality [21] and often even radical uncertainty, the firms are not normally able to optimize. Instead, they try to find satisficing solutions (based on earlier success) that can be improved by continuous learning.

There are different kinds of routines and capabilities. The simplest ones, the so-called first-order capabilities, are generated for pure replication of the existing system [20, 22]. If the environment remains stationary more than lower-order capabilities are not needed. Of course, however, normally the firms are living in continuously changing environments, which means that replication is not enough to be profitable. The firms have to generate also higher-order capabilities that are able to renew and change existing resource and knowledge bases. These higher-order capabilities are called *dynamic capabilities.* In fact, we would like to categorize capabilities as a continuum in which they range from pure replicating capabilities via semi-dynamic capabilities ("best practices") to genuine or radical dynamic capabilities that are able to generate new innovative ways to organize business activities.

In order to offer an even broader evolutionary picture we will briefly utilize the cultural evolutionary framework introduced by Campbell [23]. He distinguishes three evolutionary mechanisms that control cultural evolutionary processes, to which also business evolution belongs. The three basic mechanisms are: *variation, retention* and *selection.* In the business ecosystem, the role of *variation* is based on the firms' ability to generate something new or, as Schumpeter [12] put it, to create "new combinations" or innovations. Here the role of entrepreneurial attitude is of great importance. The second mechanism is *retention* or replication, which creates stability within the firm. Retention is typically realized by lower-order capabilities based on cumulative learning and repetitive actions following the idea of Simonian [21] "satisficing". In a way, one can think that retention is based on organizational culture. The third mechanism, *selection,* takes place through competition so that the fittest capabilities within the company and, finally, the fittest products/services, i.e., the ones that customers are willing to pay for are selected through market forces. This is the basic idea of Schumpeterian creative destruction.

In a similar way as in biological evolution in which variation is realized through mutations and retention through inheritance, it is very important that variation and retention are balanced also in the business context. To give an example, if there are too many radical innovations within a company it is most likely that an organization cannot survive because the existing organizational culture cannot cope with too many radical changes. In the biological sphere, the clear analogy is cancer as a result of two radical mutations. Interestingly, modern strategic management literature deals with this balancing problem by means of the concept of *ambidexterity*. Based on the ideas of March [24] who analyzed the roles of exploitation (based on existing capabilities, i.e., on lower-order capabilities) and exploration (based on new, not-yet-existing capabilities, i.e., on dynamic capabilities), the ambidexterity literature (see [25]) also deals with balancing these two mechanisms in a way that creates success. The more rapidly the business environment is changing the harder it is for management to balance these two tendencies, variation and retention/replication. If you invest too much in exploitation at the expense of exploration, you are not able to adjust to drastic changes in the business environment and, vice versa, if you invest too much in exploration your organization is perhaps not able to follow rapid changes due to path-dependent rigidities/organizational inertia. Clearly, also the firms in manufacturing are now facing the ambidexterity problem.

After a short evolutionary journey, we will now go deeper to look at the nature of dynamic capabilities. Dynamic capabilities were defined as follows by Teece and others [8] “a dynamic capability is the firm’s ability to integrate, build, and reconfigure internal and external competences to address rapidly changing environments”.

Another well-known definition is given by Eisenhardt and Martin [26] who stress more the “best practice” nature of dynamic capabilities thus in a way describing what we earlier called as semi-dynamic capabilities. They define dynamic capabilities as “the firm’s processes that use resources—specifically the processes to integrate, reconfigure, gain and release resources—to match and even create market change. Dynamic capabilities thus are the organizational and strategic routines by which firms achieve new resource configurations as markets emerge, collide, split, evolve, and die”. Perhaps the most exact definition is the one of Helfat and others [27], “A dynamic capability is the capacity of an organization to purposefully create, extend, or modify its resource base". In our view, this definition most clearly emphasizes Schumpeterian entrepreneurial thinking in which "new combinations" i.e. innovations are the engines of strategic (see [28, 29]).

Teece [9] went further in order to reveal the micro foundations of dynamic capabilities. According to Teece [9–11] , dynamic capabilities consist of three separate capacities i.e. *sensing, seizing*, and *reconfiguring (transformation)*. To generate real changes based on dynamic capabilities the firms have to be able to, first, sense even weak signals that appear as strategic options. Second, they have to be able to invest i.e. to exercise the most promising strategic options. Often this also means that the firms have to disinvest in some older capabilities that are not regarded as profitable any more. Third, the managers have to be able to reconfigure or transform their existent path-dependent resource and knowledge bases and processes in a way that makes it possible to realize strategic options sensed.

Again, it may happen that different capacities of dynamic capabilities are not balanced. For instance, some firms may be very effective in sensing new strategic options by means of entrepreneurial alertness and/or efficient technology scanning systems but at the same time, the managers can be quite too slow to make investment decisions thus destroying the existing opportunities. Perhaps the most problematic part is the third one i.e. how to effectively reconfigure (transform) your existing resource and knowledge bases. It is not enough to do the things right by means of operational and semi-dynamic capabilities but the managers have to do the right things through dynamic capabilities as well. Here we see the ambidexterity tradeoff problem in action. As mentioned, it is of great importance for a company to have a stable organizational culture that is based on continuous learning and exploitation of existing routines and capabilities. On the other hand, transformation necessarily means also explorative actions that destroy at least partly existing path dependent capabilities. This necessarily creates tensions on different organizational levels. There is always a tradeoff between competence-enhancing exploitation and competence-destroying exploration [25].

4 Dynamic Capabilities in Digital Platform-Based Ecosystems: How to Create and Capture Value

In the previous section, we introduced the evolutionarily inspired view of the way how firms behave and how they can achieve and sustain their CA by means of lower- and higher-level capabilities. However, during the last ten years the things have become even more complicated because of rapid globalization, keener rivalry, product and service mass customization, need of increased flexibility, modularization, digitalization and related positive network externalities. These "*digital convergence*" [3] related phenomena have dramatically changed the way how the firms nowadays compete with each other and especially how they get connected with each other through the so-called multi-sided platforms (MSP) [30]. The reason of the rapid rise of MSP's relates to positive network externalities resulting in increasing demand curves together with often dramatically decreasing average total costs due to large "first copy cost". The first mover's advantage or "winner takes it all" are the dominating principles of the MPS's. The firms that rapidly achieve the so-called critical mass are also able to establish and dominate their own platforms and thus indirectly create their own ecosystem. Because of network externalities and increased flexibility (mainly due to digitalization), the companies are able to leave their traditional industry clusters and to create quite new interfaces and ecosystems.

There are many definitions for platforms and ecosystems. We will use the ones applied in modern strategic management literature. Teece [3] defines the platforms and ecosystems as follows, "A platform is any combination of hardware and software that provides standards, interfaces, and rules that enable and allow providers of complementors to add value and interact with each other and/or users. Collectively, the platform innovator(s) and the complementors constitute an ecosystem, the viability of which depends on continued innovation and maintenance of the platform by its owner(s) and a delicate balance of cooperation and competition among the providers of complements". Helfat and Raubitschek [30] stress that "digital MSP ecosystems are characterized by crosside (or indirect) network effects, in which the value to a party on one side of the platform depends on the number and quality of the parties on the other side(s) of the platform. Cross-side network effects are often positive". *Complementary assets* play an important role in each MSP's and they are essential when a firm tries to build its business model in order to profit from its innovation. A succinct definition of the concept ecosystem is offered by Adner [31]. He regards it as "an alignment structure of the multilateral set of partners that need to interact in order for a focal value proposition to materialize". Alignment structures may be variable and the structure is often non-hierarchical consisting of complementarities that are co-specific and multi-sided. It is worth noting that partners in an ecosystem can simultaneously be also rivals.

Because of rapid digitalization, platforms and related ecosystems are becoming pervasive and therefore a provider of a complementary asset(s) has to be able to become a part of a wider platform. Even if the platforms often are multi-sided, it is typical that there is a technologically leading company that sets the agenda and controls the evolution of the platform. As Teece [32] states, an ecosystem is anchored by a platform or many rival platforms that are connected through common standards (often protected by patents, copyrights, trade secrets, etc.) and interfaces.

It is interesting to note that the idea of an ecosystems fits very well to the evolutionary theory of the firm launched in the former section. The basic idea is based on biological co-evolution, a process through which species become developed in a continuous evolutionary cycles following the variation, selection, and retention mechanisms. Moore [33] first adopted this idea in business literature in a Harvard Business Review article. Of course, Nelson and Winter [20] utilized the same idea in their evolutionary models, even if they did not use the concept of an ecosystem. As Teece [32] highlights, co-creation and co-evolution through competition (selection) are typical also for business ecosystems in which the innovator has to make transaction cost based decisions (see also [34]) about the elements of the value creating platform. The main question is which innovative ideas are to be internalized and which ones to be externalized for other providers of complementary assets.

Focusing on the importance of complementarities Jacobides and others [35] state succinctly "An ecosystem is a set of actors with varying degrees of multilateral, non-generic complementarities that are not fully hierarchically controlled". Especially they stress that the relevant complementaries can be co-specialized and unique (i.e. the items A and B cannot be produced alone without coordination that puts them together) and/or supermodular (i.e. the more an item A is produced the cheaper or better in quality are also the items B and C). Uniqueness and supermodularity mean that interdependencies are standardized, which in turn presupposes specific routines and capabilities that are needed in designing ecosystems. Helfat and Raubitschek [30] call them *integrative capabilities* and especially the ability to create them is of great importance when trying to profit from platforms and ecosystems.

In his influential Research Policy article, Teece [36] launched the idea of the profiting from innovation framework in order to analyze how a firm could profit from its innovation. Of course, the situation in the 1980s was not that complex as it is nowadays. Teece focused on one product innovation-one company-one industry-model and showed that the most important factors when trying to capture the fruits of innovation were (1) the nature of the appropriability regime (based either on tacit knowledge or on legal means, such as patents, copyrights, trademarks, and trade secrets), (2) the role of co-specialized complementary assets, (3) the nature of innovation (autonomous or systemic), and (4) timing. Appropriability regimes are strong when knowledge assets are based on tacit knowledge and, in addition, protected by legal means. Even in the 1980s, it was clear that legal means were strong only in some industries, such as pharmaceutical or chemical ones. Hence, even then the complementarities and the nature of innovation played a crucial role. Interestingly, Teece [36] in a way came to the same conclusion as the RBV at the same time. If a firm manages to have strongly protected assets that preferably are also bottlenecks in a supply chain, it is most likely able to profit from innovation as well.

In his follow-up article, Teece [3] updated his profiting from innovation framework for the digital era emphasizing the role of network externalities and digital convergence and launching the idea of an ecosystem in which the role of complementarities and multi-inventions are crucial. Teece [3] started now from platforms and ecosystems stressing the fundamental roles of complementarities and positive externalities-based interdependencies that, of course, weakened the role of the traditional appropriability regime based on legal means.

Teece [3] showed that it is hard to protect general-purpose technologies as well as so-called enabling technologies, such as photonics, advanced materials, nanotechnologies, artificial intelligence, machine learning and robotics. This means that there are too little incentives to produce new knowledge in these fields without public funding or without successfully participating in value creating ecosystems. *The key factors were the complementarities together with a strong appropriability regime.* It is not any more enough that a company is able to innovate in enabling technologies and to protect it strongly by means of tacit knowledge or legal means (cf. [37]). It has to be able to connect it to an existing or emerging digital ecosystem and, most importantly, it has to be able to generate a bottleneck asset that is unique and preferably super-modular. As Teece states [3], however, "these bottleneck assets are not easy to identify and they may shift over time" when an ecosystem evolves. A similar situation can be found from many manufacturing platforms, such like the automobiles or aircraft in which modularization and standards have made it hard to profit from autonomous innovation because of keen competition between the providers of complementarities. In fact, in the digital era competition takes place at three levels, first, between the providers of complementary assets within a platform, second, between different platforms within an ecosystem, and third, between rival ecosystems (see also [3]). In addition, all the ecosystems are continuously evolving at the same time when the interfaces are getting more blurred. Interestingly, a company can also be a part of different (even rival) platforms and ecosystems.

As Teece [3] concludes the fate of a company within an ecosystem and the fate of the whole ecosystem now drastically depends on (1) the ability to continuously generate relevant complementarities and especially on (2) the cognitive entrepreneurial capabilities [38] of the ecosystem's leaders to orchestrate, coordinate and strategize the ecosystem. In order to profit from the platforms the firms have to be able to produce bottleneck complementarities that create value for the platform and are somehow protected through tacit knowledge and timing.

This brings us to the issue of dynamic capabilities. Building up competitive platforms or bottleneck complementarities and designing business models in rapidly evolving digital ecosystems is not possible without strong dynamic capabilities. The managers of platform leading companies have to be able to sense new opportunities also outside the platforms, to seize the new opportunities rapidly if needed, and to reconfigure knowledge and resource bases not only within the own company but also within the whole platform or even ecosystem by changing its constituting elements and complements providers. In a similar way, the managers of complementary assets providers have to be able to sense, seize and reconfigure to be able to create critical bottleneck assets.

Helfat and Raubitschek [30] develop these ideas further by analyzing the dynamic capabilities that are necessary for profiting from innovation in digital multi-sided platform-based ecosystems. As they state, the platforms do not automatically generate positive multi-sided externalities but they have to be created through a deliberate design. This is mainly on the responsibility of the leader of the platform. They have to be able to orchestrate and coordinate the ecosystem consisting of many at least partly competing complementary asset providers. In addition, they have to continuously develop the "core product/service" of the platform and find the most effective complementary assets providers who are willing to join the ecosystem. This in turn presupposes the ability to balance the different needs of complementors so that they have enough incentives to be innovative. On the other hand, the leading company/companies have to be able to take their own stake. According to Helfat and Raubitschek [30] especially three types of dynamic capabilities are of vital importance for the leaders when trying to cope with multi-sided platform-based ecosystems:

1. *Innovative capabilities.* Leaders have to be able to develop the core product (product sequencing) but, in addition, they have to be able to integrate new complementors and their knowledge assets in an efficient way.
2. *Scanning/sensing capabilities.* Of course, the leaders have to sense new opportunities related to the core product(s)/ (service(s) of the ecosystem and to take into account the threats arising from the competitive environment. In addition, they have to be able to scan potential innovation sources that can be created through existing or new potential complementors. On the other hand, the complementary assets providers have to scan and sense new opportunities to make their assets as bottlenecks within the platform.
3. *Integrative capabilities* relate mainly to designing suitable business models. There are decisions about internalizing knowledge assets between the leader and complementors or between rival complementors i.e. they have to take into account the nature of governance structure based on transaction cost considerations (see [30, 34]). They also have to make decisions on the pricing structures for products/services provided within the ecosystem and between different customers. Integrative capabilities also support interactions and relationships between the members of an ecosystem as well as between ecosystem members and external parties. The more complex and knowledge-intensive interfaces there are and the more rapidly the ecosystem is evolving the more complicated is the task of orchestration. Finally, integrative capabilities are also partly responsible for how effective and innovative the sensing/scanning capabilities are.

Helfat and Raubitschek [39] launched already earlier the idea of integrative knowledge that is the basis of integrative capabilities as follows: "knowledge that integrates, or knowledge of how to integrate, different activities, capabilities, and products within a vertical chain or across vertical chains."

From the perspective of other members of an ecosystem, the role of innovative and scanning/sensing capabilities are quite similar and, of course, they need integrative capabilities, even if they do not have to take responsibility for the general governance structure of the whole ecosystem. Much more important for them is to focus on creating such a combination of especially innovative and integrative capabilities that enable them to profit from the bottleneck properties of their complementary assets. The better protected, supermodular, and co-specialized their complementary assets are the stronger their bargaining power within the ecosystem is.

5 Discussions and Managerial Implications

This article deals with obtaining and sustaining CA in manufacturing firms during the era of digital ecosystems. "Digital convergence" is drastically changing the way how the firms can profit from innovation as the examples from information technology, finance, banking, and retailing clearly manifest. In addition, the rules of the game are dramatically changing. In the future, it will not be any more possible to do everything all alone from the basic innovation to custom-tailored products and services. In the future, also the manufacturing firms have to be able to work together within digital ecosystems that often take the form of multi-sided platforms. On the one hand, these digital ecosystems are often based on positive externalities that make the systems evolve rapidly. This results in high uncertainty and the need for proactive behavior. On the other hand, these platforms and ecosystems are based on standards and very often strict modularization that does not leave very much room for individual actions. This intensifies competition between rival providers of complementary assets and makes the extra profits generated by traditional VRIN resources or autonomous innovations often temporary.

In this article, we have analyzed the nature of digital platforms and ecosystems and the way they function. In addition, we have launched the tools that economics-based strategic management literature can offer in order to achieve and sustain CA and scrutinized how effective they could be under the circumstances of digital convergence. Based on this analysis, we will now summarize our main results in the context of Industry 4.0.Transformation. It can be interpreted as a checklist that managers have to take into account when trying to profit from innovation in the digital era.

- *The Porterian message:* You have to maximize your bargaining power in relation to rivals. This can be based on the economies of scale and scope, differentiation or niche creating strategies. During the digital era most economies of scale are based on demand-related, scalable network externalities. If a company is able to exploit them it may also generate its own platform or even an ecosystem and be the leading partner within it. To be able to do this a firm needs innovative, scanning/sensing and especially integrative capabilities. It also has to be able to continuously evolve its platform and make transaction cost based internalization/externalization decisions. However, if a company is not able to generate scalable network externalities it may perhaps try to use differentiation or niche strategies but preferably as a born global trying to get internationalized as soon as possible by using the tools of Internet 4.0. Transformation. Unfortunately, local advantages based on differentiation or niches cannot be sustainable, even if a temporary CA can perhaps be achieved. This is due to keen competition within the platform and ecosystem.
- *Resource-based message:* In order to profit from VRIN resources a firm has to be able to find resources that create new value and are hard to imitate. In the manufacturing sector, the autonomous innovations are, however, hard to protect and hence the extra profits (rents) are normally only temporary. But again, if a company has dynamic capabilities (especially innovative, sensing and integrative) it can perhaps generate complementary bottleneck assets that are necessary and supermodular in nature. If a company is lucky, it can really profit from its autonomous innovations, even if there typically exists keen competition between rival providers of complementary assets.
- *Capability-based message:* First, the managers of manufacturing companies have to understand the evolutionary nature of their company and its role in platforms/ecosystems. It is not any more possible to try to survive all alone. It is necessary to function as a partner in networks and to utilize all the different capabilities from operational via semi-dynamic (best practices) to genuinely dynamic capabilities.

Most important are the ones that support innovation, scanning of new strategic options, and integration thus enabling to become a partner in platforms/ecosystems.

- The managers should also understand the message of the ambidexterity tradeoff. They have to be able to develop both their operational capability-based exploitation and dynamic capability-based exploration in a balanced way.
- Nevertheless, the more turbulent and rapidly changing the business environment is, the more the firms have to invest in dynamic capabilities. Here the basic logic of the profiting from innovation framework ([3, 36, 37]) still holds. In order to gain a CA position a firm has to be able to create (1) bottleneck complementary assets that are (2) protected either by legal means (nowadays not so important any more) or (3) tacit knowledge embedded into the organizational culture and/or (4) to utilize timing advantage. Bottleneck complementary assets are in the future more and more knowledge-based consisting of operational manufacturing capabilities, integrative and innovative dynamic capabilities as well as science-based knowledge assets. At the same time when regional wage rate cost advantages will lose their importance due to robots, additive manufacturing, etc. the companies become much more flexible and foot-loose. Our guess is that in the future, it is much more important to have close connections to the universities and research centers than try to minimize labor cost differentials. This science-based co-operation presupposes strong integrative dynamic capabilities.

References

1. D. P. Perales, F. A.Valero, and A. B. Garcia, "Industry 4.0, a classification scheme," in *Closing the gap between practice and research in industrial engineering, in Closing the gap between practice and research in industrial engineering*, edited by E. Viles, M. Ormazabal and A. Lleo, Eds., Cham, Switzerland: Springer, 2018, pp. 343–350.
2. F. Zezulka, P. Marcon, I. Vesely, I., and O. Sajdl, "Industry 4.0 – An Introduction in the phenomenon," *IFAC-PapersOnLine*, vol. 49, pp. 8–12, 2016.
3. D. Teece, "Profiting from innovation in the digital economy: Enabling technologies, standards, and licensing models in the wireless world," *Research Policy*, vol. 47, pp. 1367–1387, 2018.

4. M. Kohtamäki, M., Parida, V., P. C. Patel, P. C., and H. Gebauer, "The relationships between digitalization and servitization: The role of servitization in capturing the financial potential of digitalization". *Technological Forecasting and Social Change.* 151, February 2020, 119804.
5. M. Collan, J. Freiling, K. Kyläheiko, and E. Roemer, "Entrepreneurship and the art of tackling technological crises: a strategic real options framework", *International Journal of Technology Intelligence and Planning*, 10, 166–185.
6. B. Wernerfelt, "A resource-based view of the firm," *Strategic Management Journal,* vol. 5, pp. 171–180, 1984.
7. J. Rose, "How the U.S. can regain its manufacturing edge? An interview of Justin Rose in Knowledge@Wharton, July 10. 2019," ed, 2019.
8. D. J. Teece, G. Pisano, and A.Shuen, "Dynamic capabilities and strategic management". *Strategic Management Journal.* 18. 509–533.
9. D. J. Teece, "Explicating dynamic capabilities: the nature and microfoundations of (sustainable) enterprise performance," *Strategic Management Journal,* vol. 28, pp. 1319–1350, 2007.
10. D. Teece, "The foundations of enterprise performance: Dynamic and ordinary capabilities in an (economic) theory of firms," *Academy of Management Perspectives,* vol. 28, pp. 328–352, 2014.
11. D. Teece, "Dynamic capabilities: Routines versus entrepreneurial action," *Journal of management Studies,* vol. 49, pp. 1395–1401, 2012.
12. J. A. Schumpeter, *Capitalism, Socialism and Democracy.* New York, NY, USA: Harper & Brother, 1942.
13. M. E. Porter, *Competitive Strategy: Techniques for Analyzing Industries and Competitors.* New York: The Free Press, 1980.
14. M. E. Porter, *Competitive Advantge.* New York: The Free Press, 1985.
15. D. Ricardo, *The principles of political economy and taxation.* London, 1817.
16. R. P. Rumelt, "Towards a Strategic Theory of the Firm," in *Competitive strategic management*, R. B. Lamb, Ed., ed Englewood Cliffs, NJ, USA: Prentice-Hall, 1984, pp. 556–570.
17. J. Barney, "Organizational culture. Can it be a sound source of sustained competitive advantage?," *Academy of Management Review,* vol. 11, pp. 656–665, 1986.
18. M. Peteraf, "The cornerstones of competitive advantage: A resource-based view," *Strategic Management Journal,* vol. 14, pp. 179–191, 1993.
19. J. Barney, "Firm Resources and Sustained Competitive Advantage," *Journal of Management,* vol. 17, pp. 99–120, 1991.
20. R. R. Nelson and S. G. Winter, *An evolutionary theory of economic change.* Cambridge: Harvard University Press, 1982.
21. H. Simon, "From substantive to procedural rationality," in *Method and appraisal in economics*, S. Latsis, Ed., ed Cambridge Cambridge University Press, 1976, pp. 129–148.

22. S. G. Winter, "Understanding dynamic capabilities," *Strategic Management Journal*, vol. 24, pp. 991–995, 2003.
23. J. Campbell, "Growth Pole theory, diagraph analysis and interindustry relationships," *Journal of Economic and Social Geography*, vol. 63, pp. 79–87, 1972.
24. J. G. March, "Exploration and exploitation in organizational learning," *Organization Science*, vol. 2, pp. 71–87, 1991.
25. M. Tushman and P. Andersson, "Technological discontinuities and organizational environments," *Administrative Science Quarterly*, vol. 31, pp. 439–465, 1986.
26. K. M. Eisenhardt and J. A. Martin, "Dynamic capabilities: What are they?," *Strategic Management Journal*, vol. 21, pp. 1105–1121, 2000.
27. C. E. Helfat, S. Finkelstein, W. Mitchell, M. A. Peteraf, h. Singh H., D. J. Teece, and S. G. Winter, *Dynamic Capabilities: Understanding Strategic Change in Organizations*. Malden, MA: Blackwell Publishing, 2007.
28. F. Ardnt and L. Pierce, "The behavioral and evolutionary roots of dynamic capabilities," *Industrial and corporate change*, vol. 27, pp. 413–424, 2018.
29. O. Schilke, S. Hu, and C. E. Helfat. "Quo vadis, dynamic capabilities? A content-analytic review of the current state of knowledge and recommendations for future research". *Academy of Management Annals*, 12(1), 390–439.
30. C. Helfat and R. S. Raubitcheck, "Dynamic and integrative capabilities for profiting from innovation in digital platform-based ecosystems," *Research Policy*, vol. 47, pp. 1391–1399, 2018.
31. R. Adner, "Ecosystem as structure: An actionable construct for strategy," *Journal of Management*, vol. 43, pp. 39–58, 2017.
32. D. Teece, "Next generation competition: New concepts for understanding how innovation shapes competition and policy in the digital economy," *Journal of law, economics, and policy*, vol. 9, pp. 97–118, 2012.
33. J. F. Moore, "Predators and prey: a new ecology of competition," *Harvard business review*, vol. 71, pp. 75–86, 1993.
34. M. G. Jacobides and S. Winter, "The co-evolution of capabilities and transaction costs: Explaining the institutional structure of production," *Strategic Management Journal*, vol. 26, pp. 395–413, 2005.
35. M. G. Jacobides, C. Cennamo, and A. Gawer. "Toward a theory of ecosystems". *Strategic Management Journal*. 39, 2255–2277.
36. D. Teece, "Profiting from technological innovation: Implications for integration, collaboration, licensing and public policy," *Research Policy*, vol. 15, pp. 285–305, 1986.
37. D. Teece, "Reflections on "Profiting from innovations"," *Research Policy*, vol. 35, pp. 1131–1146, 2006.
38. C. Helfat and J. A. Martin, "Dynamic managerial capabilities: Review and assessment of managerial impact on strategic change," *Journal of Management*, vol. 41, pp. 1281–1312, 2015.
39. C. Helfat and R. S. Raubitcheck, "Product sequencing: knowledge, capabilities, and products," *Strategic Management Journal*, vol. 21, pp. 961–979, 2000.

PART III

Societal Change Brought About by Manufacturing 4.0

The Fourth Industrial Revolution and Changes to Working Life: What Supports Adult Employees in Adapting to New Technology at Work?

Hanna Nygren, Maarit Virolainen, Raija Hämäläinen, and Juhani Rautopuro

1 Introduction

This chapter aims to increase the current understanding of adults' individual learning pathways and needs, when they adapt to new technology. In this chapter, we review adults' overall technology-skills and depict, through chosen examples, how adults have adapted to the technological

H. Nygren (✉) • M. Virolainen
Finnish Institute for Educational Research, University of Jyväskylä, Jyväskylä, Finland
e-mail: hanna.nygren@samiedu.fi; maarit.ha.virolainen@jyu.fi

R. Hämäläinen
Faculty of Education and Psychology, University of Jyväskylä, Jyväskylä, Finland
e-mail: raija.h.hamalainen@jyu.fi

M. Collan, K.-E. Michelsen (eds.), *Technical, Economic and Societal Effects of Manufacturing 4.0*,
https://doi.org/10.1007/978-3-030-46103-4_10

change in their working lives. We present prior research on the challenges that the Fourth Industrial Revolution poses to adults' further education, and based on the Programme for the International Assessment of Adult Competencies (PIAAC), we review adults' problem-solving skills in technology-rich environments.

We analyse interviews of adults, who participated in initial education, to show how focal considerations, such as the usability of new technology, the economic or personal advantages of technology, and the social factors pertaining to the use of technology are influential for adults, when learning about and implementing new technology. Overall, the findings indicate the importance of design-based education and the need for companies to flexibly address adults with skill-shortages and adults who need to develop their problem-solving abilities in technology-rich environments.

The Fourth Industrial Revolution and Education

The Fourth Industrial Revolution is not limited to industrial production—rather, it is present in all fields of society and enables diverse innovation through digital technology [1, 2]. The drivers of the Fourth Industrial Revolution comprise digital, physical, and biological technologies. The revolution has changed labour markets throughout the world [3] and has set new demands for adult re-education and further education [1].

According to the OECD's *Employment Outlook* [2], roughly one-sixth of jobs in OECD countries can be easily replaced by automation, and roughly one-third may be radically transformed by technological progress. It is expected that almost half of the world's current professions will be computerised or automated to some extent [4, 5]. An estimate based on the European Skills and Jobs Survey has shown that the existing skills of the workforce in the European Union (EU) fall short roughly one-fifth of what is needed from workers to carry out their jobs at their highest productivity level [6]. In general, unemployment is more typical among those who have not reached tertiary education, and it is expected that a considerable number of jobs will be filled by those with higher education (HE) in numerous countries [2, 7].

J. Rautopuro
Finnish Institute for Educational Research, Jyväskylä, Finland
e-mail: juhani.rautopuro@jyu.fi

The assumption has been that the more repetitive and identical tasks of services and industries will be replaced with artificial intelligence and developments in manufacturing technology [3] and that automation will remove job opportunities for unskilled workers and lead to increased inequality. However, technology also creates new jobs while the old job titles and tasks are being taken over by machines and software [8]. Frank et al. [9] note that technology is typically designed to perform a specific task and thus alters the demand for specific workplace skills. At the same time, additive manufacturing contributes to the creation of jobs [10]. To take advantage of new technology, employees should be educated to collaborate with intelligent technologies in a broader sense [3]. In this context, education has been recognised as having a crucial role to play in enabling humans to adapt to the demands set by technological change. The following quote by Frey and Osborne exemplifies this:

> "While the concern over technological unemployment has so far proven to be exaggerated, the reason why human labour has prevailed relates to its ability to acquire new skills. Yet this will become increasingly challenging as new work requires a higher degree of cognitive abilities. At a time when technological change is happening even faster, a main hurdle for workers to adapt is thus the surging costs of education". [11]

Adaptation to Technology

As stated above, the Fourth Industrial Revolution sets certain demands in relation to adults' level of adaptation to technology. Adaptation is needed on societal and individual levels. Even though there cannot be a consensus as to the precise competency-demands for adaptation to technology on a task-by-task basis [9], both researchers and policymakers are unanimous in their belief that many adults need to change their profession and embed technology into their job profiles.

Meeting the competence needs created by the Fourth Industrial Revolution is challenged by somewhat interrelated obstacles. On the individual level, the development of technology skills is hampered by the low levels of participation in adult education. According to the OECD's *Employment Outlook* [2], one in five adults reported that they did not participate in training offered by their employer, because they were not sufficiently motivated, or did not find the content meaningful. It was also reported that the challenges adults faced during training reduced their interest in training [2]. The lack of participation of low-skilled adults in education and training is often due to the expense of the training, a lack of time at work or at home, or due to insufficient prerequisites to participate in training. However, our previous research provides some surprising results concerning participation in continuing education, as participation activity and problem-solving skills in technology-rich environments (TRE) seem to be unrelated [12]. The content, its relevance to the context, and the pedagogy of further education also play a crucial role in the development of problem-solving skills in TRE—participation alone does not amount to technology adaptation.

The discrepancy between adults' skill-levels in technology and the learning needs set by the changing world of work calls for societal action. Numerous governments have already taken steps to cope with this issue. For example, the EU has declared its interest in improving the innovation ability [3]. Many countries have also assembled a cast of experts to forecast technological adaptation. For example, the German Ministry of Social and Employment Affairs [13] and the Finnish Ministry of Economic Affairs and Employment [14] have published reports on the ways in which supportive policies can be implemented. Financing adult education has also been reviewed by European countries as a method of enabling learning [15]. Guoping et al. note that "the government should require and encourage enterprises to organise skill training courses and assist them to adapt to new technologies" [3]. In the case of skill shortages in relation to new technology, the governments should encourage labour-mobility between industries, and equal education rights should be ensured to promote societal equality and cohesion [3].

The adaptation to the new era of technology demands that attention be placed on how formal education and workplace learning at various enterprises can be enhanced and renewed. From the perspective of developing adult education more generally, Raivola et al. [16] have underlined the importance of offering multiple alternatives for accessing internet-based learning platforms to enable continuous participation in education and training. Furthermore, they underline that adults should be given an opportunity to complete studies that lead to acknowledged qualifications in smaller modules and that they should be credited for prior learning, work experience, and informal as well as formal learning. Finally, they emphasise that financial support should be offered to adults who are participating in further education.

The role of the workplace in providing accurate and targeted education becomes central, when industries invest in the latest technology and need their employees to adapt to technological changes. Therefore, managers should enhance staff training to enable staff members to work with intellectual technologies [3]. In this regard, three considerations influence the adoption and implementation of workplace technologies [17–19]. First, the usability of technology is a fundamental consideration—it must be natural and easy to use [20, 21]. Individuals assess the usability of technology by determining whether it increases efficiency, whether it is effective, and whether it leads to user satisfaction. The second consideration is the economic and/or personal advantage that the technology lends an individual. A threshold question when beginning to learn a new technology is whether it brings any advantage to the individual or organisation. If yes, the individual is more likely to learn how to use technology. The third consideration is the social aspects of the technology; these aspects play an essential role in the acceptance of the technology [17]. If one's peers use a specific technology, then the individual is more likely to adopt it [22].

2 Problem-Solving Skills of Adults in Technology-Rich Environments and the Demand for Design-Based Education

In this section, we detail the existing research on problem-solving skills of adults in technology rich environments (TRE) and recommend approaches for developing adult learning based on the findings.

Large-scale international comparative studies, such as the OECD's PIAAC, provide a comprehensive understanding of the varying nature of adults' skills in TRE and indicate how the skills differ between age-groups and educational backgrounds [12, 23, 24]. PIAAC provides the most pervasive overview of proficiency levels in problem-solving skills in TRE. The OECD defines problem-solving skills in TRE in the following way: "Problem-solving in technology-rich environments involves using digital technology, communication tools and networks to acquire and evaluate information, communicate with others and perform practical tasks" [25]. The first PIAAC problem-solving survey focused on "the abilities to solve problems for personal, work and civic purposes by setting up appropriate goals and plans, accessing and making use of information through computers and computer networks" [25].

In the PIAAC study, the scale of problem-solving skills was divided into four proficiency levels. On average, more than one-third of participants did not reach Level 1 [26], meaning that they did not take the computer-based test because they only had some or no previous experience of computers. The first proficiency level out of four differentiated groups was formed of those who did not reach Level 1. Level 1 was reached by those who were able to use familiar technological applications such as e-mail and browsers. The tasks at this level did not demand navigation or only did so to a minor extent, and they involved completing just a few of tasks. At Level 2, the tasks required advanced abilities in relation to more specific technologies, where participants had to navigate and integrate information. At Level 3, both the generic and specific use of technologies and inferential reasoning were required.

The findings from the PIAAC study [12, 23, 24] have highlighted that problem-solving skills in TRE are unevenly distributed among adults and that there is a need to develop new approaches to adult education. The problem-solving skills of adults with vocational education and training (VET) backgrounds in TRE are weaker than the problem-solving skills of adults who have completed HE. At the same time, for those adults with HE and strong problem-solving skills in TRE, having a skilled occupation led to them having stronger problem-solving skills [24].

Further exploration of the PIAAC data has indicated that the association between formal learning and problem-solving skills in TRE seems to be quite weak and that informal learning activities are heavily associated with capable problem-solving skills in TRE. Thus, design-based learning activities should be provided and studied through interventions and case studies [27]. Accordingly, in line with the PIAAC findings, researchers have suggested implementing tutoring programmes and scaffolding at work to support the enhanced adoption of technological skills [24]. Adults should be provided with group-based working and learning approaches that support problem-solving skills in TRE. They note that adults' non-formal and informal learning as well as the development of their professional competencies, knowledge, and skills should be supported to enable them to respond to future workplace needs [23].

3 Reflections and Experiences from Adults on the Adaptation to Technology at Work

In the following three subsections, we reflect on the experiences of the participants of this study, who are four adult learners ($N = 4$) who found employment through technology-based training that was organised by a company who adopted new technology in metal processing. The company had difficulties in finding employees who were competent with this new technology. Thus, there was a special interest in organising training to meet the recruitment needs of the company. The training was organised in cooperation with the local vocational school.

The cases of the four participants are presented here to obtain a more detailed understanding of the experiences of adults who are already vocationally educated, have successfully completed the training demanded by their company, and have begun to work with technology. Based on the examples drawn from the analysed cases, we depict the main influential factors in relation to the adaptation and implementation of new technology. They are (1) the usability of the technology; (2) the economic and/or personal advantage the technology poses to an individual; and (3) the social factors that support adaptation to the technology [17, 20, 21]. We then conclude with the main viewpoints that should be considered when helping adults learn and adapt to changes in their working life due to the developments of Industry 4.0.

The participants were selected for training based on their applications to participate in the training offered by the company. Their initial training lasted three months. The adults spent two days a week at the local vocational school, which organised the training. The rest of the week was spent at work practicing what they had learned in training. The training at the local vocational school consisted of learning the basics about metals, work safety, first aid, LEAN basic training as well as reading graphic designs, practicing welding, and using the bending machine. During the training, the adults did not receive any salary from the company. As a form of financial compensation, however, they received unemployment benefit or a study grant from the government or trade union.

The participants ($N = 4$), three men and one woman, were aged between 20 and 55. They were interviewed at their workplace by the first author, and the topics of the thematic interviews comprised their previous experience with technology, how they make use of it at work on a daily basis and during their leisure time, how technology was related to their social lives, what kind of benefits they saw in it, and how they viewed the usability of technology during their free time and while at work. Each of the participants are referred to here using pseudonyms.

The participants already had some previous work experience with metal, welding, and machines in general and they had an interest in technology. In the following paragraphs, we clarify, using detailed examples from the interviews, how they saw the usability of technology, the economic and personal advantages it provides, and the social factors that support adaptation.

The Usability of Technology

In this subsection, we explore the participants' views on the usability of technology at work and during their free time. The participants were asked to estimate how much technology is needed in their current role. It was notable that they *positively* described their adaptation to the technology used at work and that they found it usable in several ways. In the following two examples (see quotes from Anni and Tomi) describe what their daily work consists of, how technology makes their work faster, and how it increases their productivity, accuracy, and safety:

Anni: *You need them* [referring to technology] *a lot, for example, these programmes are running, and you see them on the computer. You can read the pictures and you are able to zoom in on the screen. You can send messages to your colleagues and ask "What is this and is this done right or wrong?" You can send a message with a picture to your boss and say, "Hey, I did this, is it okay" You can't say that you could work without it* [technological skill], *because the pictures are made with a computer; you draw, measure. It would be more challenging and difficult without the computer.*

Tomi: *It is important* [referring to technological skills] … *You need to choose the right programme. Otherwise, you can spoil the piece if you don't have the right programme.*

The participants describe the many ways they find technology meaningful, motivating, and emotionally satisfactory. They also reflect on their personal professional development and their transcendence of their earlier competence boundaries. When the participants were asked about their expectations when they began technology training, Anni reported that she wanted to learn to use the technologies in a fluent way, while Mika expected to get a permanent contract at work (Example 4):

Anni: *Mmm ... That I would learn something that I could apply also in my free time and I would learn to understand why a product is like this, how it is made, what is it made of.*

Mika: *To get a permanent job in what I have now and well ... well, well, and that the work would be as I had thought. Everything has hit the spot. Workmates are nice and stuff ... I have liked it. There are no problems.*

When the participants were asked to reflect on themselves as learners of technology, Tomi and Mika were strong in their expressions. They seemed to be motivated and excited to begin something new and had made a committed decision to learn.

Tomi: *Like, whatever comes, I'll take the bull by the horns. I'll just learn new things. It is about yourself. You will learn it, or you will not.*

Mika: *Well, I just decided that I will learn this; it helped. I was so interested in this job, and I was also interested in learning. I put a lot of effort into this. And it worked! At least on some level.*

Combining learning from work experience with professional education has been modelled to take place best through a combination of theoretical, practical and self-regulated knowledge in the integrative pedagogical (IP) model [28]. In recent elaborations on the (IP) model, the importance of paying attention to emotions of a learner has been highlighted. The IP model has been used internationally to design learning environments and combine learning at school and at the workplace. The examples show how, in addition to enabling the integration of theoretical, practical, and self-regulative knowledge in learning environments, the meaning of emotions related to self, context, tasks, and performance should be understood [29–31].

Economic and Personal Advantages of Adapting to Technology

In this subsection, we highlight how determining the economic and personal advantages of technology utilisation promotes its adoption [17]. The participants were asked how they viewed their future employment opportunities in the field of production (where they were working). They anticipated the employment as good and stable (see quote from Anni). They appeared to be satisfied with their decision to participate in re-training.

Anni: *You can work with a variety of things. Metal is the thing that people like. Interior design is a good example, you see a lot of metal* [used in interior design] *at the moment. You can design something that you never believed you could until you see it by yourself. The future seems quite bright.*

The participants also described how they had enjoyed the personal advantages of technology through information retrieval and communication with their friends and family. Anni also mentioned the convenience of being able to take care of one's own business on a constant basis, being able to follow the most up-to-date news, and the ease of being able to conduct online banking (see quotes from Anni):

Anni: *Well, sending messages, basic banking things; you don't have to queue at the bank to pay your bills. You see your account information and you get real-time information. You don't need to wait many days for a newspaper.*

Anni: *... An app for a doctor's office to make an appointment ... And then there is the veterinarian's FirstBeat app, if something happens to your dog. You get a direct connection through the internet, and you don't need to worry about what is wrong.*

These examples show how the participants have been able to find meaningful ways of using technology in their private and professional lives, thus lowering their barriers in relation to new technology. They had positive experiences with the usability of the new technology at home and at work. It has been found that individual values, personal history, and engagement in other social practices are central definers of participation in workplace practices and learning through work [32, 33]. The practice of commanding work-related assignments by professionals in the workplace is ´situated, dynamic, founded in and relational to the practice´ where it is embedded [33]. Thus, merely paying attention to the affordances of learning, i.e. what kind of cues and learning opportunities are provided by supervisors and colleagues or through the guiding manuals and programmes as learning materials is insufficient. In addition, the learners should have opportunities to explore the meaning of what they have learnt, to reflect on their experiences and to construct their understanding about the technology, themselves as its users and the meaning of the technology for the broader context of the workplace and their social context. However, the utilisation of technology may be selective and based on personal preferences, as the following example (quote from Mika) shows:

Mika: *I have watched a few videos on YouTube and practiced afterwards.*

In contrast, Tomi stated that although he owns a computer and a couple of tablets, he is lazy and lacks the motivation to use them at home, where he mostly uses technology for communication and for retrieving information. Even though he used technology on a daily basis at work, it was not as important at home. He found the old-fashioned way of taking care of things during his leisure time to be preferable and found comfort in not having to put extra effort into the old ways of doing things. In sum, his use of technology at work did not predict his use of technology at home:

Tomi: *Well, quite a bit, I use it myself … I don't know what it is … I'm kind of an old-fashioned man. In the market, I prefer to pay for my shopping using a real person, and the post office is the same: I prefer to go through the cashier than use the automaton.*

The Social Factors That Support Adaptation to Technology

The following examples show how social factors intervene when learning new technology. Social relations can afford informal technology learning, but they can also restrict it. The participants were asked to describe their use of social media and other communication technologies. The participants used these technologies mostly because they wanted to stay connected with their friends and families, many of whom live far away (see quote from Anni):

Anni: *I think it's useful when you have friends and you can send messages asking how they are and stuff. My parents live 400km from my home, so calling them and sending messages is important. You maintain your relationships.*

The participants were also asked where they got help with technology issues if they needed it. Anni and Mika reported that they got help from their friends, family members, and colleagues (see quotes from Anni and Mika):

Anni: *Friends. Mmm … Well, friends are those who help you … when they know a lot of stuff and they have the education and they know computers.*

Mika: *I get help from my friends and workmates.*

When the participants were asked about the social support they received when applying for and beginning their training, they mentioned they support they received from their families. When they were at work, they were supported by their co-workers in their training (see quotes from Mika):

Mika: *Family. Family largely. Mom and Dad.*

Mika: *Well … the management in some way, and then the workmates, and … everybody says that they can help, and everyone has been nice to me. No one is against me. You get along with everyone.*

The final example (see quote from Martti) shown here is typical of the contradictory attitude in general towards technology. Even though Martti agrees that technology is very useful and that its applications can be widely used in everyday life, he is not satisfied with it. He uses technology at work but says that he does not want to use it at home:

Martti: *Technology is very useful. You can take care of things 24/7. You are not tied to time. And then there are the robots in manufacturing, and in medicine, and even the cars work with technology. But me, I am very poor with technology. At home, we have the scribe* [referring to his wife]. *She does everything because she knows it better. Even though I know that I could and would learn if I wanted to … I am not interested. It is a necessary evil.*

4 Discussion

The Fourth Industrial Revolution is making demands for education and training, especially on those who have to adapt to technology and re-educate themselves. The participants in this study already had a background in vocational education and training qualifications before participating in the training provided and demanded by their company. According to the results of the PIAAC, VET-educated adults are more likely to have insufficient problem-solving skills in TRE [25]. In the

OECD *Employment Outlook* [2], the respondents reported that they were not sufficiently motivated or did not find the content sufficiently meaningful to participate in training to enhance their technology skills.

The interviews conducted in this study reveal the individual pathways that adults are taking to adapt to technology at work. In contrast to the results of the PIAAC and the *Employment Outlook*, the participants in this study were motivated and positive regarding their adaptation to technology. The participants described their daily work routines and how technology has increased the productivity, accuracy, and safety of their work [20, 21]. As the examples show, they also believed that their working life was stable and that the rate of employment in their field was good. In addition, they stated that technology was useful for their leisure time [17]. As well as this, social factors played a role in their adaptation to technology. All of the participants stated that they used messenger apps and social media to communicate with their family and friends, although the level of use differed among the participants. They all acknowledged the usability of these communication tools. They also stated that they got help from their friends and family in relation to any technology issues they had [17, 22]. Nygren et al. [27] note that adults' informal learning activities are highly associated with sufficient problem-solving skills in TRE. Everyday life learning and skills used outside the workplace are clearly related to sufficient skills in TRE [12, 27].

To foster the adaptation of technology, three elements should be considered when designing training programmes for adult technology learners. These elements are usability, the economic and/or personal advantage it bestows, and the social factors that support its adaptation [17–19]. Based on the reflections of the participants, it appears that these elements assist technological learning. According to the OECD *Employment Outlook*, "A comprehensive adult learning strategy is needed to face the challenges of a changing world of work and to ensure that all workers, particularly the most vulnerable, have adequate opportunities for retraining throughout their careers" [2]. Adults who need to adapt to technology should be provided with design-based, safe, and supportive learning activities [27]. It should also be noted that technological skills and competencies are also learned in formal and informal contexts. As presented in this study, new approaches to adult education [12, 23, 24] should ease adults' adaptation to technology and respond to the educational needs of the Fourth Industrial Revolution.

References

1. Schwab, K. (2016). *The Fourth Industrial Revolution*. New York, NY: Crown Business.
2. Organisation for Economic Co-operation and Development (OECD). (2019). *OECD employment outlook: The future of work*. Paris: OECD Publishing. DOI: https://doi.org/10.1787/9ee00155-en
3. Guoping, L., Yun, H., & Aizhi, W. (2017). Fourth Industrial Revolution: Technological drivers, impacts and coping methods. *Chinese Geographical Science, 27*(4), 626–637. DOI: https://doi.org/10.1007/s11769-017-0890-x
4. De Vos, A., Dujardin, J. M., Gielens, T., & Meyers, C. (2016). *Developing sustainable careers across the lifespan: European Social Fund Network on Career and AGE (Age, Generations, Experience)*. Springer International PU.
5. Frey, C. B., & Osborne, M. (2013). The future of employment: How susceptible are jobs to computerisation? *Oxford Martin School*. Retrieved January 15, 2014, from http://www.oxfordmartin.ox.ac.uk/downloads/academic/The_Future_of_Employment.pdf
6. Cedefop. (2018a). *Insights into skill shortages and skill mismatch: Learning from Cedefop's European skills and jobs survey*. Luxembourg: Publications Office. DOI: https://doi.org/10.2801/645011
7. Cedefop. (2018b). How many jobs will be taken by those with medium/high-level qualifications? *Cedefop*. Retrieved August 10, 2019, from https://www.cedefop.europa.eu/pt/publications-and-resources/statistics-and-indicators/statistics-and-graphs/36-how-many-jobs-will-be
8. Acemoglu, D., & Restrepo, P. (2018). The race between man and machine: Implications of technology for growth, factor shares, and employment. *American Economic Review, 108*(6), 1488–1542. DOI:https://doi.org/10.1257/aer.20160696
9. Frank, M. R., Autor, D., Bessen, J. E., Brynjolfsson, E., Cebrian, M., Deming, D. J., … Rahwan, I. (2019). Toward understanding the impact of artificial intelligence on labor. *Proceedings of the National Academy of Sciences, 116*(14), 6531–6539.
10. Kianian, B., Tavassoli, S., & Larsson, T. C. (2015). The role of additive manufacturing technology in job creation: An exploratory case study of suppliers of additive manufacturing in Sweden. *Procedia CIRP, 26*, 93–98. DOI: https://doi.org/10.1016/j.procir.2014.07.109
11. Frey, C. A., & Osborne, M. A. (2015). Technology at work: The future of innovation and employment. *Oxford Martin School*. Retrieved July 30, 2019, from http://www.oxfordmartin.ox.ac.uk/downloads/reports/Citi_GPS_Technology_Work.pdf
12. Hämäläinen, R., De Wever, B., Malin, A., & Cincinnato, S. (2015). Education and working life: VET adults' problem-solving skills in technology-rich environments. *Computers and Education, 88*(October), 38–47.

13. Hoppmann, N., Huber, M., Mikfeld, B., Neufeind, M. P., Primavesi, A., Rahner, S., ... Zirden H. (Eds.) (2017). *Arbeit weiterdenken.* [Thinking about the future of work] Werkheft 03. Berlin: Federal Ministry of Labour and Social Affairs. Retrieved October 11, 2019, from https://www.arbeitenviernull.de/fileadmin/Downloads/BMAS-Werkheft-3.pdf
14. Finnish Ministry of Economic Affairs and Employment. (2019). *Edelläkävijänä tekoälyaikaan* [*Leading the way into the era of artificial intelligence*]. Helsinki: Työ- ja elinkeinoministeriö.
15. Cedefop. (2019). Financing adult learning database. *Cedefop.* Retrieved August 10, 2019, from https://www.cedefop.europa.eu/en/publications-and-resources/tools/financing-adult-learning-db/overview.
16. Raivola, R., Heikkinen, A., Kauppi, A., Nuotio, P., Oulasvirta, L., Rinne, R., Kamppi, P. & Silvennoinen, H. (2006). Aikuisten opiskelumahdollisuudet ja järjestäjäverkko toisen asteen ammatillisessa koulutuksessa. Koulutuksen arviointineuvoston julkaisuja 15. [Adults'educational opportunities and the network of education providers for the upper secondary vocational education]. Jyväskylä: Koulutuksen arviointineuvosto, Koulutuksen arviointisihteeristö.
17. Coovert, M. D., & Thompson, L. F. (2014). *The psychology of workplace technology.* New York, NY: Routledge.
18. DeLone, W. H., & McLean, E. R. (1992). Information systems success: The quest for the dependent variable. *Information Systems Research, 3*(1), 60–95.
19. DeLone, W. H., McLean, E. R., & Smith, G. E. (2003). The DeLone and McLean model of information systems success: A ten-year update. *Journal of Management Information Systems, 19*(4), 9–30.
20. Gillan, D. J., & Bias, R. G. (2014). Usability science II: Measurement. In M. D. Coovert, & L.F. Thompson (Eds.), *The psychology of workplace technology* (pp. 162–182). New York, NY: Routledge.
21. Hancock, P. A. (2014). Human factors. In M.D. Coovert, & L.F. Thompson, (Eds.), *The psychology of workplace technology* (pp. 149–161). New York, NY: Routledge.
22. Cascio, W. F., & Montealegre, R. (2016). How technology is changing work and organizations. *Annual Review of Organizational Psychology and Organizational Behavior, 3*(1), 349–375.
23. Hämäläinen, R., De Wever, B., Nissinen, K., & Cincinnato, S. (2017). Understanding adults' strong problem-solving skills based on PIAAC. *Journal of Workplace Learning, 29*(7/8), 537–553. DOI:https://doi.org/10.1108/JWL-05-2016-0032
24. Hämäläinen, R., De Wever, B., Nissinen, K., & Cincinnato, S. (2019). What makes the difference: PIAAC as a resource for understanding the problem-solving skills of Europe's higher-education adults. *Computers and Education, 129*, 27–36.

25. Organisation for Economic Co-operation and Development (OECD). (2016). *Skills matter: Further results from the Survey of Adult Skills.* Paris: OECD Publishing. DOI: https://doi.org/10.1787/9789264258051-en
26. Friberg, T., Rosdahl, A., Halapuu, V., Valk, A., Malin, A., Hämäläinen, R., … Mellander, E. (2015). *Adult skills in the Nordic region: Key information-processing skills among adults in the Nordic region.* Copenhagen: Nordic Council of Ministers.
27. Nygren, H., Nissinen, K., Hämäläinen, R., & De Wever, B. (2019). Lifelong learning: Formal, non-formal, and informal learning in the context of problem-solving skills in technology-rich environments. *The British Journal of Educational Technology, 50*(4), 1759–1770. DOI: https://doi.org/10.1111/bjet.12807
28. Tynjälä, P., Häkkinen, P., & Hämäläinen, R. (2014). TEL@work: Toward integration of theory and practice. *British Journal of Educational Technology, 45*(6), 990–1000.
29. Arpiainen, R. L., Lackéus, M., Täks, M., & Tynjälä, P. (2013). The sources and dynamics of emotions in entrepreneurship education. *Trames Journal of the Humanities and Social Sciences, 17*(67/62), 331–346.
30. Tynjälä, P., Kallio, E., & Heikkinen, H. (2020). Professional expertise, integrative thinking, wisdom and phronesis. In E. Kallio (Ed.), *Development of adult thinking: Perspectives from psychology, education and human resources.* Milton Park, UK: Routledge (in press).
31. Hökkä, P. K., Vähäsantanen, K., Paloniemi, S., & Eteläpelto, A. (2017). The reciprocal relationship between emotions and agency in the workplace. In M. Goller & S. Paloniemi (Eds.), *Agency at work. Professional and Practice-based Learning,* (pp. 161–181). Cham: Springer.
32. Billett, S. (2001a). Learning through work: Workplace affordances and individual engagement. *Journal of Workplace Learning, 13*(5/6), 209–214.
33. Billett, S. (2001b). Knowing in practice: Re-conceptualising vocational expertise. *Learning and Instruction, 11*(6), 431–452.

Modelling the Societal Division of Added Value Created Through Manufacturing 4.0

Heikki Hiilamo and Henri Aaltonen

1 Introduction

Ever since the seminal study by Frey and Osbourne, scholars have engaged in a lively debate on the future of work [1–3]. In this debate, the primary issue discussed has been the expected impact of technological change, which includes broad and vague concepts, such as automation, robotization, ever increasing computing power, Big Data, the penetration of the Internet, the Internet-of-Things, online platforms and artificial intelligence [3–5]. In the context of this chapter, we summarize technological changes under the term Manufacturing 4.0. Irrespective of the term used, one school of thought claims that machines will displace human labour, and not just blue-collar but also white-collar tasks, which will consequently result in major labour-market disruptions [6]. This dystopian vision calls for social policies to protect people and societies from the devastating effects of mass unemployment, although scholars disagree on the scope and rate of change in the labour market. Another school of thought

H. Hiilamo (✉) • H. Aaltonen
University of Helsinki, Helsinki, Finland
e-mail: heikki.hiilamo@helsinki.fi; henri.aaltonen@helsinki.fi

M. Collan, K.-E. Michelsen (eds.), *Technical, Economic and Societal Effects of Manufacturing 4.0*,
https://doi.org/10.1007/978-3-030-46103-4_11

emphasizes job polarization, both in terms of wages and employment vulnerability, between routine middle-skilled workers, on the one hand, and low-skilled and high-skilled non-routine workers, on the other. While Manufacturing 4.0 will create benefits and wealth for leading global manufacturing companies that employ automated and robotized manufacturing systems, the replacement of human workers with automation and the geographical shift of manufacturing will potentially create at least temporary waves of unemployment in the manufacturing sector [7].

Under these circumstances, a clear need exists to create systems that allow companies to enjoy the fruits of their investments in new technologies while ensuring the ability of industrial society to cope with these changes. As the networked business models related to Manufacturing 4.0 open a global game, where the location of the tax domicile of a given company becomes a point of optimization, industrial societies may wish to create structures that incentivize rather than discourage the establishment of Manufacturing 4.0 companies. The question of how to find meaningful employment for those at the margins of the labour market already is already of immediate relevance throughout the developed world. However, the debate on the effects of Manufacturing 4.0 has focused mainly on the role of technology, while impacts on other sectors of society, such as societal institutions, including social protection, have gained less attention [8, 9]. Nevertheless, different models of social policy adaptation are already being discussed around the globe [10]. This chapter discusses social policy adaptations that incentivize rather than discourage Manufacturing 4.0 and other technological-change driven disruptions while ensuring the normal functioning of the industrial society.

It is important to note that Manufacturing 4.0 not only threatens social protection in the welfare states. These new technologies also provide welfare states with new untapped opportunities to deliver social security, such as easily updatable income registers and personal social accounts. In addition, it should be recognized that new technologies may enable production near the workforce and sources of raw materials without large-scale investments in the production systems. Larger production entities can be realized through the networking of small and mid-sized local manufacturers, which may reinvigorate semi-urban environments, as local manufacturing maintains and develops current service businesses and creates demand for new services.

This chapter studies the social policy responses to the perceived challenges of Manufacturing 4.0 in three countries with strong manufacturing and/or transportation industries and current low unemployment, namely Germany, the Netherlands, and South Korea. These countries were also chosen, because they have experienced lively debates on the effects of Manufacturing 4.0. Two of the countries, namely the Netherlands and Germany are welfare states and they belong to the European Union. Meanwhile, South Korea is a liberal market economy and it is a member of the Association of Southeast Asian Nations.

The analysis is based on expert interviews focusing on four themes: (1) emerging needs and solutions for social protection and employment promotion in the new labour markets of 2030, (2) possibilities to enable production near the workforce and sources of raw materials without large investments in equipment, (3) new ways of delivering social security utilizing new technologies and (4) new innovations to limit the possible labour market disruptions caused by automatization (e.g., basic income).

We begin by describing the context of the debate in the three countries. We then present the methodology and the data. The analysis section is divided into four subsections: early signs of labour-market disruptions, the role of semi-urban environments, new ways of delivering social security, and basic income and other innovations to reform social security. Finally, we discuss our key findings and the ways in which they can contribute to the wider debate on Manufacturing 4.0 and social security.

2 Economy and Social Contexts

Traditionally, Germany, the Netherlands and South Korea have possessed strong manufacturing industries. The German manufacturing industry includes the automotive sector, chemicals, metals such as iron and steel, electrical equipment, coal, ships, machine tools, high precision equipment, optics, pharmaceuticals, textiles, and plastic goods [11]. In the Netherlands, the key industrial areas consist of agriculture, the chemical industry, the creative industry, the high-tech industry, energy, and manufacturing industries such as metallurgy, electronics, and life science [12]. The most dominant South Korean manufacturing industries are steel, car manufacturing, shipbuilding, and electronics. South Korea is the world's largest producer of semiconductors.

In 2018, Germany was ranked fifth in the world in terms of nominal GDP; moreover, it enjoyed a GDP growth rate of 1.2%. By comparison, in the same year, the Netherlands was ranked 17th for nominal GDP and experienced a GDP growth rate of 1.5%, while South Korea achieved the 14th highest nominal GDP and a 2.5% GDP growth rate [13].

Germany, in addition, has one of the lowest inflation rates, at around 0.5%. Furthermore, in 2018 it enjoyed an unemployment rate of just 4.2%. Its labour force is composed of 45 million people, of whom 73.8% are engaged in the service sector, 24.6% in industry and 1.6% in the agricultural sector [14]. The Netherlands, in turn, has an inflation rate of 2.5% (2018) and an unemployment rate of 4% (2017). The majority of workers are employed in the service sector (80.6% in 2015), followed by industry (17.2% in 2015) and agriculture (1.2% in 2015) [14]. For South Korea the corresponding figure for inflation is 1.9% (2017) and for unemployment 3.8% (2018). In terms of the labour force, 70.6% work in the service sector, 26% in industry (2017) and 4.8% in agriculture (2017) [14]. Figure 1 illustrates the unemployment rates in the three countries.

The automotive industry is one of the main driving forces behind Germany's accelerating economy, and the country is the primary location for innovative car manufacturers and suppliers. Moreover, with the constant growth in aircraft production, aerospace opportunities in Germany seem promising. Furthermore, the chemicals industry, which is one of the largest and oldest industries in Germany, plays an elemental role in the

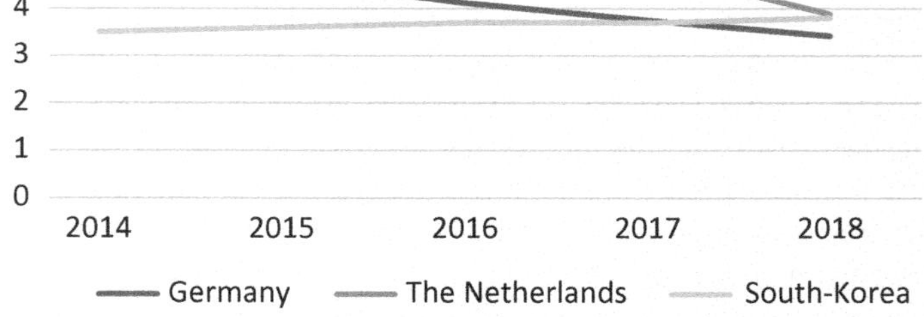

Fig. 1 Unemployment rate in Germany, Netherlands and South Korea between the years 2014–2018 [14]. (Data Source: theglobaleconomy.com (accessed 23 September 2019))

country's economy. The key elements of its success include new materials, energy efficient processes, and solution-driven engineering expertise [15].

The German electronics industry is the largest European manufacturer of semiconductors and displays, and the "Silicon Saxony" region is one of the top five semiconductor clusters worldwide. Moreover, Germany's ICT market is the largest in Europe [15]. However, Germany has experienced much discussion about slow digitalization and the immense difficulties of adopting it. Indeed Germany lags significantly behind rest of Europe in digital infrastructure [16]. In the German case, the most salient question is how the country can expand and speed-up its digital infrastructure. At the time of writing (2019), many parts of Germany lack a functioning and reliable digital network. This lack of digital infrastructure hinders economic development and makes Germany an odd country in Europe, with its huge digitally barren tracks of land.

As the world's fifth-largest exporter of goods, the Netherlands occupies a prominent position when it comes to world trade. In 2015, the Netherlands exported goods worth a total of almost USD 668 billion, approximately 3.5% of total global exports. The Netherlands is also a significant exporter of commercial services—exports of commercial services amounted to USD 189 billion (€ 160 billion) in 2016, placing the country sixth in world rankings. Interestingly, the Netherlands is also the world's second largest exporter of agricultural and food-products after the USA. As well as being a major exporter, the Netherlands also imports large quantities of goods: US$ 507 billion in 2015. Accounting for a 3.1% share of total global imports, the country is the world's eighth-largest importer of goods [17]. Moreover, the Netherlands can be considered a forerunner in adopting to digitalization. Digital technologies—such as big data analysis, artificial intelligence, blockchain, 3D printing, cloud storage and computing, and the Internet of Things—are being used in an increasing number of fields [18].

South Korea, in turn, has experienced remarkable success in combining rapid economic growth with significant reductions in poverty. Per capita income increased from USD 100 in 1963 to almost USD 30,000 in 2018. After two years of economic stagnation, GDP growth in South Korea rose slightly to 3.1% in 2017 and 2.8% in 2018, due to a rebound in household consumption, improvement of the real estate sector and fiscal and monetary stimulus measures. Weakening exports and stagnant investment, however, along with a failure to translate the boom in the chip sector into growth in other industries, are projected to limit economic growth to

2.6% and 2.8% in 2019 and 2020 respectively [19]. Moreover, although the unemployment rate has been decreasing (3.7% according to IMF and 3.4% in December 2018, according to Statistics Korea) the number of irregular workers is extremely high, social inequalities are deepening and social ties are deteriorating [20].

The South Korean government is making a concerted effort to improve the digital skills of the country's workforce. A 2016 report by Barclays jointly ranked South Korea and Estonia first in the world in terms of their ability to equip their workers for the digital economy [21]. This report stated that "Estonia and South Korea are joint leaders in digital empowerment" [21]. It lauded South Korea's world-leading track record in broadband provision in addition to the well-balanced nature and effective implementation of its digital skills policy in formal education as well as adult training, including those out of work. Moreover, the South Korean government has invested heavily in R&D, focusing on technology tied to the fourth industrial revolution.

As Table 1 demonstrates, the three countries are fairly similar in terms of poverty and inequality, with the Netherlands enjoying the lowest levels of both poverty and the GINI-Index.

3 Methodology

The methodology of this research was qualitative [23]. On the basis of earlier research, we identified four main themes to be investigated. They include signs of disruptions in labour markets caused by Manufacturing 4.0, the role of semi-urban environments in the adoption of Manufacturing

Table 1 Poverty and inequality in Germany, the Netherlands, and South Korea in 2018 [22]

Country	*Poverty rate with 60% median income poverty threshold*	*GINI index*	*Dataset*
Germany	16.7	0.293	DE15 (2015)
Netherlands	12.4	0.264	NL13 (2013)
South Korea	20.1	0.306	KR12 (2012)

Data source: Luxembourg Income Study, lisdatacenter.org (accessed 24.10. 2019)

4.0, new ways of delivering social security utilizing new technologies, and new innovations to limit possible labour-market disruptions caused by automation (e.g., basic income). We then developed a set of questions, which addressed the four major themes relevant for modelling the societal division of added value created through manufacturing 4.0. The selected experts were presented with the questions presented below:

1. Digital innovation's ability to enable machines to accomplish an increasing number of complex tasks is signalling changes to the world of work. Flexible labour markers require flexible social security systems. Can you foresee disruptions in your country's labour markets caused by automation? Even weak signals?
2. New technologies enable production near the workforce and the source of raw materials without large investments in equipment, which may reinvigorate semi-urban environments. Is this something which is visible in your country?
3. Automation, digitalization and artificial intelligence are not only threats to social protection and the welfare state. They also provide social security systems with new untapped opportunities, such as easily updatable income registers and personal social accounts to deliver social security. Which new ways of delivering social security have been discussed in your country?
4. Basic income has been proposed as a solution to labour-market disruptions caused by automation. In which contexts has basic income been discussed in your country? Have any other new innovations been proposed to reform social security (e.g. personal accounts, participation income)?

We used case-studies to analyse the three different countries (Germany, the Netherlands and Korea). In a case-study setting, a well-placed sample can be used to explain a much larger context [24]. Case-studies also offer the methodological possibility to make generalizations from a small group of cases. Thus, even a small sample allows us to compare the differences between the cases and to draw conclusions.

Among a relatively small group of informants, individual experts are in pivotal position within the specific context of their country. To find information on the themes presented above, we identified academic experts who had participated in the debate on Manufacturing 4.0 from universities in Germany, the Netherlands and South Korea. We used Google scholar to find research articles and Google to find blogs and conference presentations on adaption to Manufacturing 4.0. Once an expert had been identified, we also used snowballing to identify more experts.

Only a limited number of experts in each country (see Table 2) agreed to participate in the study. After several rounds of requests, we were able to recruit three experts in Germany, one expert in the Netherlands and five experts in South Korea. All the experts were sent the questions presented above. The answers were collected through personal interviews from experts in the Netherlands and South Korea. By contrast, in the case

Table 2 Experts who participated in the study

Name	*Country*	*University/ Affiliation*	*Academic position*	*Description of competence with relevance for Manufacturing 4.0*
Martin Gross	Germany	University of Tübingen	Professor	Labour market and social inequality
Aljoscha Jacobi	Germany	Humboldt University, Berlin	Assistant professor of sociology	Labour market and digitalization
Philip Staab	Germany	Humboldt University, Berlin	Professor of sociology	Digitalization, labour and industry sociology
Minna van Gerven	Netherlands	University of Twente	Professor	Welfare states globally
Hyosang Ahn	South Korea	Basic Income Earth Network in Korea	Executive director	Basic income and student activism
Young Jun Choi	South Korea	Yonsei University, Seoul	Professor for public administration	Planned basic income experiment for the city of Seoul
Iljin Hong	South Korea	Yonsei University, Seoul	Research professor	Comparative welfare state research
Nam Hoon Kang	South Korea	Hanshin University, Seoul	Professor of economics	Chair of Basic Income Earth Network in Korea
Jun Koo	South Korea	Korea University, Seoul	Professor of economics, Dept. of Public Administration	Innovation and public policies, planned basic income experiment for the city of Seoul

of the German experts, the questions were sent and the answers received via email. The language used in the questions and interviews was English, and all but one expert answered in English (one German expert answered in German).

Once the data were collected, we analysed the answers using qualitative content analysis.

4 Analysis

In the following, we summarize the experts' opinions on signs of labour-market disruptions, the role of semi-urban environments, new ways to deliver social security, and basic income and other new innovations for social security.

Signs of Disruptions

Our informants were able to identify early signs of labour-market disruptions, but there was no major loss of jobs in sight in any country. Germany was the country with the weakest signs of disruption. Automation has been a German success story, and consequently German industry has been quick to adapt to new technological innovations. Moreover, the German experts viewed the future as more of a gradual shift than a rapid transformation to Manufacturing 4.0. For example, the manufacturing of automobiles is slowly moving towards the production of electric cars. As one expert put it, *"[F]orecasts are always perilous, but if I had to make a bet, I would say that we most likely will see more gradual task-specific change, which will have a strong negative impact only on people with some level of specialized qualifications performing automatable tasks."* The focal point in Germany concerns the regulation of platform work, as a divide exists between those whose labour is regulated and those whose labour is relatively free from regulatory control. The question in Germany, as well in the other countries in our sample, is how to identify vulnerable groups and develop measures to include them in future labour markets.

In the Netherlands, few signs exist of drastic labour-market disruptions. The country's unemployment rate has been one of the lowest in the Eurozone for a decade. Moreover, the Netherlands enjoys good digital infrastructure, and its strong position as an export country and its highly educated workforce make the country very able to capitalize on the opportunities created by digitalization. For instance, the Netherlands has traditionally featured an extremely strong logistics sector, and the country boasts many cloud-based logistics and innovation-hubs. Dutch society is currently embracing the economic and social opportunities created by digitalization, as well as leveraging the opportunities it has for the public sector, such as for health care, mobility, energy and the food sector. Moreover, signs of sudden disruption have yet to be seen.

The Dutch social security system places a strong emphasis on flexibility. This flexibility allows companies the necessary room for manoeuvre in a small and open economy such as the Netherlands. Moreover, it also helps citizens combine work and family responsibilities without major interventions from the public sector. This flexibility is manifested primarily in the predominance of part time work, which is very common in the Netherlands, especially among women. Part-time work, however, comes with strong social protection and thus cannot be defined as atypical work or involuntary work, as it is in many other European countries. In some sectors, temporary work is in high demand (particularly in the agrisector), and the number of people with temporary contracts (instead of permanent contracts) has been growing across the Dutch labour market. Furthermore, self-employment is also becoming more common in the Netherlands. Together with the rise of platform economies, self-employment has been emphasized as a solution to the challenges of future labour markets in the country.

In essence, the Dutch welfare state is no longer a welfare state but a participation state. The Dutch system expects every citizen to be active and find work in the labour market or engage in some other form of participation outside the labour market. While the degree of participation differs, the aim is activate the unemployed and prepare them for working life. In the Dutch system, municipalities enjoy a great degree of autonomy and thereby the independence to find their own solutions to participation, which can differ greatly from municipality to municipality. The Dutch system emphasizes self-reliance, and if this is insufficient, the family is expected to provide assistance. Nonetheless, if individuals lack the resources to help themselves, the municipality, and in the last resort the state, will provide support.

The Korean experts, in turn, observed that the risk of job losses due automatization was significantly lower in South Korea than in many other countries because of high levels of expertise in the labour force. Korea has already experienced extensive robotization in the manufacturing industry.

Moreover, robotization and automatization are generally considered a positive change for Korea.

Nevertheless, as large companies dominate Korean industry, SMEs are less adapted to automatization and offering quality jobs. Signs also exist of a growth in unemployment through automatization. Thus, while the Korean economy is growing, this growth is not reflected in a rise in employment. Consequently, contrary to previous growth periods, an accumulation of value is not mirrored in an accumulation of jobs. In addition, service sector jobs are simultaneously disappearing. However, the country's extremely low fertility rate and small number of migrants are reducing the size of the labour force, which helps Korea adapt to the loss of jobs resulting from Manufacturing 4.0.

Minimum income, in turn, has risen in recent years because automation increases the price of human labour. One expert summarized the situation thus: "*Minimum income has been increased considerably over the last few years. That incentivize[s] Manufacturing 4.0 since [the] price of human labour increase[s]. Actually [the] minimum income rise has benefitted middle and high income earners whose salaries have increased but has not helped low-income workers who have experienced more unemployment.*"

Role of Semi-urban Environments

To date, few signs exist of reinvigorated semi-urban environments in Germany. Moreover, the core urban environment is experiencing high pressure and structural change. Germany currently faces a shortage of skilled labour in many areas, which, in turn, accentuates urbanization. Start-ups are primarily founded in larger cities to attract talent and benefit from cluster-effects. Some developments in the areas surrounding larger cities are evident, but they can be considered part of normal economic development.

Similarly, in the Netherlands, innovations and market developments tend to occur in large cities and triple-helix networks (public, private and knowledge institutions). No stringent segregation of the labour market between urban and rural areas exists in a country as densely populated as the Netherlands. However, in regions such as Zeeland, Friesland and Limburg, job opportunities are scarcer and the population age is higher than in the capital area (Randstad) and other *brainport areas* (university areas).

Korea, in turn, is characterized by a dualized labour market, which means it is polarized between large companies and SMEs. Put simply, large companies are investing in new technologies, including robotization, while SMEs are not. Moreover decentralization has occurred. For example, the South Korean city of Unsal already enjoys a higher productivity level than Seoul.

Korea considers itself a global centre of manufacturing, which makes manufacturing 4.0 of paramount importance to the Korean national economy. To maintain a competitive environment for manufacturing companies, Korea continues to invest in robotization and automation. That means robots are replacing human labour, with robot hubs located around existing cities or rising in entirely new locations. As one expert put it, the development *"is characterized by a gradual decline in employment by new technology rather than a change in location by new technology*".

The Korean population is aging even faster than that of Japan, which means that provincial cities are declining and the population is compressed into large cities. This trend has led to the desolation of rural areas and provinces. Rapid aging also significantly affects welfare, as the life expectancy of Koreans is the highest among OECD countries. This is due to universal health insurance and a vegetable-based diet. As a consequence of Korea's aging population, the dependency ratio is bound to increase.

Rapid aging also causes generational conflict, which is set to intensify in the future. Thus, while pensions can be guaranteed for the older generation, the situation for the following generations remains uncertain. In the words of one expert, "*It is becoming more difficult to establish generational justice with existing traditional pension schemes. The burden for young people is getting bigger. This is why basic income is needed*".

New Ways to Deliver Social Security

The debate concerning new social security systems in Germany is dominated by the themes of flexibilization and, as in the 1980s, the possible reduction of working time. The German debate is characterized by two different approaches to social innovations. The first emphasizes the universalizing of existing social policies, which tend to be employment-centred; i.e. they exclude the self-employed and the economically inactive. Examples include the universalization of public health insurance or minimum pensions. The second approach aims at innovation within employment-centred policies. The Social Democratic Party (SPD), in particular, has proposed several ideas. One has been the model of "Arbeitzseitkonten", which would grant employees the opportunity to benefit from the difference between calculated working hours and actual working hours, with the difference to be used later at a time chosen by the employee [25]. In turn, "Lebenschancekredit" [26] is a system designed to give employees the freedom to choose how and when to use their labour hours. Lebenschancekredit is basically a bank account where one store excess working hours. In other words, both schemes provide workers with unconditional access to periods of free time with the right to return to the same, safe position (of which there are far fewer than 40 years ago). Thus, flexible work and flexible free time are at the core of the German discussion.

In Germany, attitudes towards the digitalization of public administration services remain rather sceptical. However, some local administrations have digitized basic public services, such as applications for passports, and drivers' licenses. Interestingly, when the central government was overwhelmed by the task of registering Syrian war refugees and processing their cases, it relied on a mix of expanding services and digitizing the relevant government agency. This progressed to the extent that AI was used to classify refugees. Nevertheless, this move led to a backlash, and thus if applied on a larger scale to all citizens, it would be politically costly. In addition, some steps have been taken towards digitization in public health insurance, but to not in other major federal-level social programmes.

In turn, the digitalization of society is a buzz word in the Dutch debate. Digital government and the digitalisation of public administration have been important goals for recent governments. These include digitalization of the administration of social security. Moreover, a number of e-democracy tools have been made available in the last decade to increase citizen participation and fuel civil-society platform-technology thinking. Moreover, while trade unions are concerned about the effects of digitalization and automatiion on the labour market, they are reasonably positive about digitalization itself and the economic and social opportunities that it offers. Thus, the Dutch approach could be characterized as civil-society platform-technology thinking. Nonetheless, trade unions remain undecided on which opinion they will adopt vis-a-vis digitalization. Furthermore, Dutch society is waiting to see the kind of stance Brussels will take on the matter before reacting and advancing pragmatically.

By contrast, the central theme in the social security debate in South Korea concerns selectiveness in the social security system. Selectiveness is a more costly and hence less effective way to deliver social security than the universalist approach. There is a hope, however, that new technologies will reduce screening costs. The remaining caveat is that a reduction in screening costs does not necessarily mean that the basic needs of the vulnerable are met. Currently, automation is not utilized extensively in delivering social security. There is, however, a strong consensus around increasing coverage and spending in social security. If and when this occurs, new technology can be applied.

Basic Income

The debate on basic income is at an early stage in Germany. However, a recent shift has occurred from intellectual debates to political debates. Nonetheless, concrete measures are still missing. While basic income is a rather popular concept, which is regularly discussed in the media, and which, according to opinion polls, is supported by roughly half of Germans, the experts participating in our study saw no clear consensus emerging. Moreover, proponents of basic income support it for different and often contradictory reasons, suggesting that any specific model would have weaker approval ratings. Furthermore, support for basic income only weakly aligns with social class or party preferences, making it harder to mobilize voters around the topic.

Two parties, the Greens and SPD, have models of basic income that focus on improving the current social safety net of last resort. One expert described the situation with the following remark: "*One SPD model departed widely from the original idea of basic income by demanding that recipients do 'socially beneficial work', i.e. the term 'basic income' was propped [up by] the older concept of a 'social labour market', where the state provides low-skilled jobs for the long-term unemployed*". In other words, two strains of thought are found in Germany; the first emphasizes relieving pressure on the state budget, while the second stresses quality of life and the ability to decide the kind of life one wishes to lead.

In the Netherlands, by contrast, municipalities are currently experimenting with basic income. These trials focus on current welfare claimants (social assistance) and therefore go beyond mere 'basic income'. The process has been colourful: The experiments were intended to be launched on May 2017, but they were subsequently delayed. They finally began in June 2018, and are set to conclude in October 2019. The trails aim to provide an alternative approach to social assistance and investigate the effects of a different set of rules on claimants of social assistance. In the political debate within the participation-state modus, the experiments are seen as way to increase active participation in labour markets and support those who cannot find work. The current experiments employ different approaches, as Dutch municipalities differ greatly in the ways they enforce participation: whereas some municipalities employ very strict conditions and sanctions, others are far more lenient. At the time of the writing, the results of these recent experiments are not yet known.

The basic income movement in South Korea began from grass-roots societal dissatisfaction. Support for basic income stems from young people's perception of a lack of solidarity among Korean citizens. The basic income debate is, consequently, mostly preoccupied with the situation of young people in society. However, while the city of Seoul planned a basic income experiment for young people, the initiative failed to gain sufficient support. In Korea there are two opposing views on the development of the social security system, a welfarist approach and basic-income approach. Currently, the welfarist perspective, which emphasizes work ethics and the prevention of free riding, remains stronger.

However, Gyenoggi province, with 13 million inhabitants, has begun a basic income programme where all young people will receive 1 million won (around 760 euros) at the age of 24 for one year without any conditions. The Gyenoggi basic income is paid in a local currency which is valid only in this province. Moreover, only SMEs can convert this currency into legal currency. Thus, the aim is both to help young people and support local SMEs. Moreover, this is considered one of the largest pilots with local currency in the world.

In Korea, two additional variations of basic income also exist. One is basic income as land dividend: a policy of imposing a land tax and distributing the income equally among the people. The second variation is targeted at farmers, and is, in effect, participation income. Here, the main aim is to improve the situation of poor farmers.

Gyenoggi's basic income for young people plays a pivotal role in the basic income debate in South Korea. As one expert noted, "*The SMEs' business opportunities are less important that what is happening to you[ng] people, changes in young people's attitudes, economic aspects and study choices*". If basic income effects a positive change, more young people will move to Gyenoggi province from other parts of the country, and this will significantly increase pressure to implement basic income at a national level.

5 Discussion

The industrial structures and path-dependencies in their social security systems make each of the three countries unique cases. Nonetheless, it is still possible to find similarities between the chosen countries. The analysis above demonstrates a clear need for social policy adaptation to Manufacturing 4.0 in countries with highly developed manufacturing industries and current low unemployment levels. Despite the diverse nature of the concept of social security and the myriad social security schemes in existence around the world, all three countries recognized the necessity to reform their social security systems in the wake of prospective labour-market disruptions. The question is if whether these reforms will be sufficient and when the reforms are needed.

In the German case, no labour-market disruptions were foreseen in the near future because of the robust level of automatization in the country's various industrial sectors. Small and medium-size enterprises (SMEs or *Mittelstand*) are the backbone of German industry and the source of larger companies' economic might. Nevertheless, the focal point for Germany, as well for the other countries in this study, concerns the regulation of platform work. There is a gap between those whose labour is strictly regulated and those whose labour is more loosely controlled. Digitalization is progressing swiftly for large companies, but many SMEs are struggling in this area. Due to the German federal system, municipalities are in a key position to enhance digital infrastructure, whereas federal bodies play a central role in advancing the path of digitalization.

In the Dutch case, no signs are evident of severe labour-market disruptions. Moreover, this finding is of no surprise. The Netherland's unemployment rate has been one of the lowest in the Eurozone for a decade. Furthermore, the country possesses good digital infrastructure, and its strong position as export country with a highly educated workforce make the Netherlands more than able to capitalize on the opportunities created by digitalization. Thus, the Netherlands has all the ingredients to succeed in digitalization.

In South Korea, the risk of job losses due automation is significantly lower than in many other countries because of the high levels of expertise in the labour force and the decrease in labour supply caused by the country's extremely low birth rate. Korea already enjoys a high level of robotization in the manufacturing industry. Moreover, robotization and automation are generally considered a positive change for Korea.

Apart from the need to renew their social security systems in response to technological change, all three countries face an ageing population, thus making it harder to finance social security schemes. The countries have all faced various challenges in their past, which they have overcome, but in different ways. This has led to a different perception of how society should deal with unemployment and the renewal of social security schemes.

To date, few signs exist of reinvigorated semi-urban environments in Germany. Furthermore, the country's core urban environment is experiencing high pressure and structural change. Germany currently faces a shortage of skilled labour in many areas, thus larger cities attract much of the potential labour force, enhancing their position to the detriment of smaller towns and rural areas. Similarly, in the Netherlands, innovations and market developments occur in large cities and triple-helix networks (public, private and knowledge institutions). The Netherlands is nonetheless a densely populated country, and consequently urban and rural areas exist in close proximity. However, some areas lag behind the capital area in their pace of change.

The population is ageing in South Korea at a record pace, even faster than that of Japan. Consequently, provincial cities are declining and the population is increasingly centred in large cities. This is leading to the desolation of rural areas and provinces.

The debate concerning new social security systems in Germany is dominated by the themes of flexibilization and, as in the 1980s, the possible reduction of working time. The German debate is characterized by two different approaches to social innovations. The first emphasizes the universalization of existing social policies, which tend to be employment-centred; i.e. they exclude the self-employed and the economically inactive. The second approach aims at innovation within employment-centred policies. The SPD, in particular, has proposed several ideas to promote inclusive labour markets. In terms of attitudes towards the digitization of public administration services, the climate remains fairly sceptical in Germany. While some steps towards digitization have been taken in public health insurance, other major federal-level social programs have failed to follow suit. This encapsulates the attitude towards new solutions for social security and how one can control the data behind these innovations.

In turn, the buzz word in the Dutch debate is the Digitalization of society. Thus, recent governments have stressed the goals of Digital government and the digitalization of public administration. The Dutch approach could be described as a civil-society platform-technology thinking. However, before reacting more forcefully, Dutch society is waiting to see the stance Brussels adopts on digitalization. This summarizes the current Dutch attitude towards digitalization.

The central theme in the social security debate in South Korea, by contrast, concerns selectiveness in the social security system. Selectiveness is costly, and hence represents a less effective means of delivering social security compared to the universalist approach. There is a hope, however, that new technologies will reduce screening costs. Nonetheless, a strong consensus is forming around increasing social security coverage and spending. If and when this transpires, new technology can be applied. Korean society is focused on cost and the efficiency, which are tied to the digitalization of social security.

The three countries' social security systems have been forged over long periods of time in different historical contexts, and this is also reflected in their approach to basic income. The debate on basic income is at an early stage in Germany. However, as previously mentioned, a recent shift has occurred from intellectual debates to political debates, although concrete measures are still lacking. Moreover, while basic income is regularly discussed in the media and, according to opinion polls, enjoys the support of roughly half of Germans, the experts in our study saw no signs of a consensus emerging. Thus, although Germans are pondering the feasibility of basic income for their society, there is little indication of concrete action in the foreseeable future.

As mentioned earlier, the Netherlands is currently experimenting with basic income at a municipal level. These trials focus on current welfare claimants (social assistance) and therefore go beyond mere 'basic income'. In the political debate within the participation-state modus, the experiments are seen as a means of increasing active participation in the labour market and supporting those unable to find work. The current experiments employ different approaches, as Dutch municipalities differ greatly in the ways they enforce participation, with some municipalities employing very strict conditions and sanctions and others being far less stringent. At the time of the writing, the results of these recent experiments are as yet known. Regional diversity is one peculiarity of the Dutch system, where every municipality approaches participation as it sees fit. Nonetheless, participation in working life is the essence in the Dutch welfare state, which can be considered a participation state.

The basic income movement in South Korea began from grass-roots societal dissatisfaction. Support for basic income stems from young people's perception of a lack of solidarity among Korean citizens. The basic income debate is, consequently, mostly concerned with the situation of young people in society. This quest for more solidarity is the driver of changes in young Koreans' participation in the labour market. Nonetheless, the Korean labour market is dominated by large conglomerates, and consequently little room exists for innovative, original ideas. However, Korea features two additional variations of basic income. One is basic income as land dividend: a policy of imposing a land tax and distributing the income equally among the people. The second variation is targeted at farmers, and is, in effect, participation income. Here, the main aim is to improve the situation of poor farmers.

Reliability and Validity

We were only able to recruit a limited number of experts to act as informants for this study. This is unsurprising given that the topic is new and research knowledge has only recently begun to accumulate. The data were collected by both personal interviews and email. However, the same questions were used for both data collection methods. While the selected experts do not constitute a representative sample of all existing expert opinions in their respective countries, given their background and experience we assume that they were able to convey valid information on the topics analysed in this study. Our qualitative research method allowed us to study nuances and create a more coherent picture of the themes than that available in the current literature.

6 Conclusion

To conclude, in Germany the main challenge is society's perception that unemployment is primarily a personal issue. Work is abundant, and thus one must merely desire to work. Similarly, the Netherlands is a participation state, where the responsibility for finding work or other means of activation lies with the individual. If a person is unable to cope, help should be sought from the family, then from the municipality and, as a last resort, from the state. By contrast, in South Korea social innovations tend to be locally-driven ad-hoc campaigns, thus making their future somewhat bleak. Consequently, Korea seems to lack an overarching strategy for applying basic income or other social innovations. Only time will tell whether the approaches discussed in this study will provide working solutions for citizens struggling to find work and relying on social security schemes and whether the three countries will successfully navigate the paradigm shift to Manufacturing 4.0.

References

1. C. E. Frey and M. Osborne, "The Future of Employment: How Susceptible Are Jobs to Computerisation?" Oxford Martin Programme on Technology and Employment, Sept. 17, 2013.
2. C. E. Frey and M. Osborne, "The Future of Employment: How Susceptible Are Jobs to Computerisation?" *Technological Forecasting and Social Change,* vol. 114(C), pp. 254–280, 2017.
3. D. Acemoglu and P. Restrepo, "The Race between Man and Machine: Implications of Technology for Growth, Factor Shares, and Employment," *American Economic Review*, vol. 108, no. 6, pp. 1488–1542, 2018.
4. M. Arntz, T. Gregory and U. Zierahn, "The Risk of Automation for Jobs in OECD Countries: A Comparative Analysis," OECD Social, Employment and Migration., OECD Publishing. Paris, no. 189, 2016.
5. D. Autor, "Why Are There Still So Many Jobs? The History and Future of Workplace Automation," *Journal of Economic Perspectives,* vol. 29, no. 3, pp. 3–30. 2015.
6. A. Korinek and J. Stiglitz, "Artificial intelligence and its implications for income distribution and unemployment," NBER series, Dec. 2017.
7. J. Cowie and J. Heathcott, "The meanings of deindustrialization," in *Beyond the ruins: The meanings of deindustrialization*, J. Cowie and J. Heathcott, Eds., ed Ithaca, NY, USA: ILR Press, 2003, pp. 1–15.
8. D. Acemoglu and J. Robinson, *Why Nations Fail: The Origins of Power, Prosperity and Poverty.* London, England: Profile Books, 2012.
9. D. Acemoglu and J. Robinson, *The narrow corridor. States, societies and the fate of liberty.* New York, USA: Penguin Press, 2019.
10. H. Hiilamo and K. Komp, "The case for participation income: acknowledging and valuing the diversity of social participation," *The political quarterly,* vol. 89, pp. 256–261, 2018.
11. B. E. Sawe, "The Economy of Germany," *WorldAtlas*, April 2017, [Online]. Available: worldatlas.com/articles/the-economy-of-germany.html. Accessed: September 20, 2019.
12. Netherlands Enterprise Agency, "Key Sectors" *Holland Trade and Invest* [Online]. Available: https://www.hollandtradeandinvest.com/key-sectors. Accessed: September 20, 2019.
13. Gross domestic product (GDP). doi: 10.1787/dc2f7aec-en[Online]. Accessed September 23, 2019.

14. Global Economy, [Online]. Available: https://www.theglobaleconomy.com Accessed: September 23, 2019.
15. German Trade & Invest, "Industries in Germany" [Online]. Available: https://www.gtai.de/GTAI/Navigation/EN/Invest/industries.html. Accessed October 24, 2019]
16. K. Schwab, *The Fourth Industrial Revolution*. Geneva, Switzerland: World Economic Forum, 2016.
17. Netherlands Foreign Investment Agency, "Economic Overview" [Online]. Available: https://investinholland.com/economic-overview/. Accessed: September 20, 2019.
18. Goverment of the Netherlands, "Dutch digitalisation strategy" June 1, 2018, [Online]. Available: https://www.government.nl/documents/reports/2018/06/01/dutch-digitalisation-strategy. Accessed: September 20, 2019.
19. Nordea, "South Korea: Economic and Political Overview" February 2020*, [Online]. https://www.nordeatrade.com/fi/explore-new-market/south-korea/economical-context. Accessed: September 20, 2019.
20. C. Kim and S. Lee, "South Korea proposes record budget lifeline for jobs, welfare in 2019" August 28, 2018 [Online]. Available: https://www.reuters.com/article/us-southkorea-economy-budget/south-korea-proposes-record-budget-lifeline-for-jobs-welfare-in-2019-idUSKCN1LD02Y. Accessed: October 24, 2019.
21. Barclays, "From Inclusion to Empowerment: The Barclays Digital Development Index" July, A 2016 [Online]. Available: https://www.longitude.co.uk/wp-content/uploads/2018/11/Barclays-Digital-Development-Index.pdf. Accessed: October 24, 2019.
22. Luxembourg Income Study Database (LIS) [Online]. Available https://www.lisdatacenter.org/ Accessed: October 24, 2019.
23. J. Jupille, "Knowing Europe, Metatheory and Methodology in European Union Studies" In *Palgrave Advances in European Union Studies*, M. Cini and A. Bourne Ed., London, England: Palgrave Macmillan, 2006, pp. 209–232.
24. B. Flyvbjerg, "Five Misunderstandings About Case-Study Research," *Qualitative Inquiry*, vol. 12, no. 2, pp. 219–245, 2006.
25. Prosoft, "Arbeitszeitkonto, Was ist ein Arbeitszeitkonto?" [Online]. https://www.prosoft.net/was-ist/arbeitszeitkonto. Accessed: October 24, 2019.
26. S. Mau, "Der Lebenschancenkredit: Ein Modell der Ziehungsrechte für Bildung, Zeitsouveränität und die Absicherung sozialer Risiken," WISO, Friedrich-Ebert-Stiftung, Bonn, 2015.

Coping with Technological Changes: Regional and National Preparedness in Face of Technical Change

Jari Kaivo-oja, Mikkel Stein Knudsen, and Theresa Lauraeus

1 Introduction

Countries around the world face the dual challenge presented by economic pressure and technological change. Our home country, Finland, can be seen as a canary in the coalmine for the need to meet new dynamics of increasingly volatile, complex, and ambiguous (VUCA, [1]) conditions. Finland was hit particularly hard by the Great Recession, and in 2009 alone Finland lost 8.5% of its GDP [2]. The workforce employed in industrial manufacturing was reduced by about 25% from 2007 to 2016, losing about 90,000 jobs during this period [3]. However, while certain national, economic, and geopolitical conditions exacerbated the direness of Finland's situation, the challenge of coping with technological changes is

J. Kaivo-oja • M. S. Knudsen (✉) • T. Lauraeus
Finland Futures Research Centre, University of Turku, Turku, Finland
e-mail: Jari.Kaivo-oja@utu.fi; Mikkel.Knudsen@utu.fi; Theresa.Lauraeus@utu.fi

M. Collan, K.-E. Michelsen (eds.), *Technical, Economic and Societal Effects of Manufacturing 4.0*,
https://doi.org/10.1007/978-3-030-46103-4_12

a global one. How do we measure, and how do we ensure regional and national preparedness in the face of the technical changes provided by the Fourth Industrial Revolution and Industry 4.0?

It should come as no surprise that not all nations or regions are equally prepared for this challenge. In terms of preparedness for Industry 4.0, evidence from patent data on advanced manufacturing processes and systems showed that many of Europe's capabilities in the field were based on a small group of nations [4]. In many regions, and indeed in entire Member States, such capabilities appeared outright absent.

Given the enthusiasm with which policymakers and key business and civil society representatives joined into 'surf the Industry 4.0 wave' during the latest years, surprisingly little empirical literature exists on the links between Manufacturing 4.0 and national specialization [5]. While a number of firm-level Industry 4.0 maturity models and readiness analyses have been developed [6, 7], models are still only emerging for societal levels of analysis. Lobova et al. [8] assesses levels of countries' formation of Industry 4.0 by measuring (1) the level of society's digitization, (2) mentions of Industry 4.0 in normative and legal governments of the state, and (3) total volume of financing of scientific research in the sphere of Industry 4.0. Sung [9] ranks *competitiveness for the fourth industrial revolution* by averaging three global competitiveness analyses.

This chapter argues that neither of these concepts fully capture regional and national abilities to cope with relevant technological changes. Instead, measurements of preparedness must consist of several elements including both technology push-factors (supply) and technology pull-factors (demand). As elaborated during this chapter, the supply side represents the existing industrial strengths of spatial entities, as well as the ongoing technological innovation. The demand side represents changes in future markets, fundamental changes to business models, and elements of technology foresight. To put it in simple terms, countries succeeding in the future industrial landscape will be those able—also in the future—to match their own competences with national and international demands.

In Fig. 1 we have presented supply (push) and demand-side (pull) challenges of Industry 4.0 strategy. In the field of technology innovation management, there is always a strategic need to analyze both supply and demand aspects of technological innovation. Successful technological innovation requires that both sides of market interaction (supply and demand) are managed successfully and simultaneously (see Fig. 1).

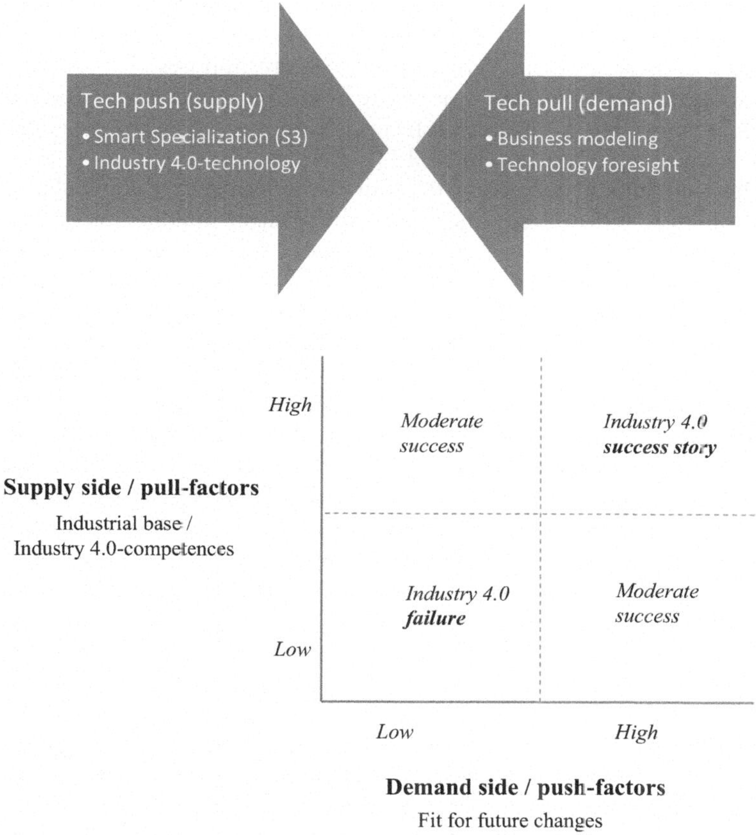

Fig. 1 Supply (push) and demand-side (pull) factors in the Industry 4.0 strategy

Our approach builds on a modified version of the concept of Smart Specialization in which we integrate a stronger futures focused perspective supplementing the evolution and refinement this theoretical concept has witnessed over the last decade. Smart Specialization Strategies (S3) is a knowledge-driven growth strategy, which remains the dominant answer of the European Union for the regional transformation challenges faced. In 2016 European Commissioner for Regional and Urban Policy Corina

Cretu termed smart specialization a '*key instrument for place-based development*' and '*the most comprehensive policy experience of implementing innovation-driven progress in Europe*' [10]. S3 presupposes that choices of technological domains should originate in prior analyses of a region's economic, technological, and research specializations in order to be aimed at those with the greatest potential [11].

This chapter is structured by the four elements in Fig. 1 above. First, S3 is explained in more detail, since this is the primary building-block of current policymaking on this topic, as well as one of the dominant academic paradigms in the field. In order to succeed in the new conditions, proper attention must also be given to assessing and developing technical competences in the field of Industry 4.0. Since other chapters of this book focus more on Industry 4.0-technologies, this chapter will intentionally keep this topic short. On the pull/demand-side, the two core elements of the model relate to new modes of business modeling and the need for technology foresight. New business models and new markets are paramount to keeping a region competitive. The final part of the chapter demonstrates the benefits of foresight and applications of futures research methods. Throughout the chapter, we will use examples and cases based in Finland, but the concepts and ideas are transferable and easily applicable in other countries.

Smart Specialization (S3)

As mentioned, Smart Specialization has become the main adaptational mantra of the European Union. Smart Specialization, or S3 for Smart Specialization Strategy, ensures the capacity of an economic system (a region for example) to generate new specialties through the discovery of new domains of opportunity and the local concentration and agglomeration of resources and competencies in these domains [12]. Smart Specialization, more so than traditional industrial policy, places great emphasis on geographical contexts including the existing social, cultural, and institutional characteristics [13]. In S3 as a sector-specific innovation policy, governments (national, regional) are reframed with the task of developing capabilities connected with its territory, without accusations of merely 'picking the winners' in selecting fashionable or desirable sectors in a manner disjointed from existing specializations [14]. There are various reasons why S3 strategies are different in different European countries:

- The location and importance of industrial, export and import activities are different in different regions of national economies.
- The key elements of the S3 approach, industrial comparative advantages, regional resilience levels, and innovation activities are different in different regions.
- The industrial scale and scope factors are different in different regions.
- The nature of regional collaboration and actor-network network patterns are different in different regions inside national economies.
- Both private- and public sector decision-makers need a tailored regional knowledge base and information assets for investment decisions.
- The Entrepreneurial Discovery Process (EDP) and start-up ecosystems are different in different locations and regions. The nature of start-up ecosystems is having regional special characters.
- In general, pre-conditions of Industry 4.0/Manufacturing 4.0 strategy concerning Industry 1.0–3.0 developments, are different inside national economies, in regions.

Smart Specialization Strategy (S3) is an excellent example of a theoretical concept translated into policy. In S3, the recommendation is that resources should be concentrated on a limited number of well-defined priorities. The S3 approach thus requires tough choices on the basis of own strengths and international specialization [15]. The aim is not to create more specialized—and thereby less diversified—economies, but to capitalize on and leverage existing strengths in a region in order to build a competitive advantage in high-value activities [16]. S3 thus defines 'the virtuous process of diversification' [17] through local concentration of resources and capabilities within a select number of domains particularly likely to generate new activities aiming at transforming the existing structures, it forms capabilities by building micro-systems of innovation and drives structural local changes. The selected S3 priorities should be based on a shared vision built during a wide consultation process, which should be socially inclusive. It should include a wide range of entrepreneurs, researchers, social partners, etc. Priority setting should rely on the logic of the entrepreneurial discovery process (EDP) of likely market opportunities [18].

The place-based tenets of S3 are primarily aimed at underpinning regional development at a level more fine-grained than the national levels. The bottom-up approach reflects a popular turn from national to regional innovation systems over the latest decades [5]. Part of this reason is the empirically verified benefits from the size and critical mass in R&D combined with the positive knowledge spillovers gained from proximity to sufficiently large R&D sectors within a given sector [12]. Since the spillover varies with geographical proximity, there is a tendency to aim at innovation ecosystems sufficiently large to create size, but not larger than that.

There are tendencies that fundamental changes to the manufacturing business model might underpin the regional spillover even further. As the line between manufacturing and services is blurred, emergent literature on servitization suggests, "Territorial Servitization" can contribute to local competitiveness and employment. This happens through a virtuous cycle generated when a local manufacturing base attracts or stimulates the creation of complementary knowledge-intensive service businesses, which in turn may facilitate the creation of new manufactures in the region [19, 20].

Smart Specialisation potentials across Finnish regions have been extensively evaluated in cooperation with the Manufacturing 4.0-project in Haukioja et al. [21] and Karppinen et al. [22]. In the context of preparedness for impacts of technical changes, we must understand the socio-economic historical background of regions. Some regions are more ready for technological changes and Industry 4.0 transformations. Key variables like demographic structure and education level of the population have impacts on preparedness for impacts of technological changes.

Like with any concept gaining rapid importance, many competing descriptions of Smart Specialization-elements exists. This chapter focuses on four main elements needed for a Smart Specialization-theory (visualized in Fig 2).

We want to underline that in the S3 approach the entrepreneurial discovery processes (EDPs) always link to the development of spatial start-up ecosystems. Entrepreneurial discovery always encourages experimentation and risk-taking. Some of the new economic activities identified as priorities are likely to fail. Otherwise, one could hardly speak of experimentation. This aspect is also relevant for innovative technological entrepreneurship and Industry4.0 / Manufacturing 4.0 developments.

Below we will expand further on the four main constitutive elements of Smart Specialization. We can claim that these four constitutive elements of Smart Specialisation Strategy (S3) promote the creation of Industry 3 and Industry 4.0-type manufacturing patterns in regions. These must be some comparative advantages of industrial activities in order to reach the Industry 4.0 level in different industrial sectors. In the European Union,

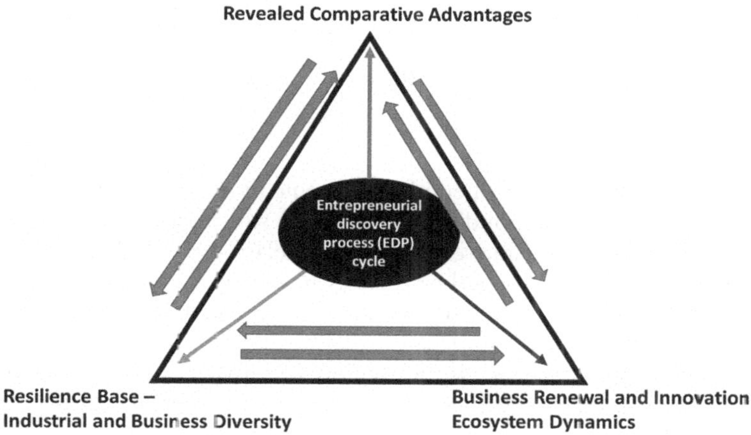

Fig. 2 S3 triangle (comparative advantage, resilience, and business renewal innovation ecosystem dynamics) with entrepreneurial discovery process

the existence of a national strategy for Smart Specialisation Strategy (S3) has been an ex-ante conditionality for the use of the European Union Structural Funds from 2014–2020 [23]. This means that if private and public stakeholders want to link Industry 4.0 development and investments to the use of structural funds of the EU, they should have a very good knowledge base of S3 smart industrial development in regions.

Revealed Comparative Advantages

A new systemic approach to Smart Specialisation Strategy has been developed in the publications of the S3-research group [24, 21]. The revealed comparative advantage (RCA) is a key concept in S3 strategy, as regions seek to upgrade along with their existing advantages and specializations. Typically, the revealed comparative advantage is defined to be an index used in international economics for calculating the relative advantage or disadvantage of a certain country/region in a certain class of goods or services as evidenced by trade flows or industrial activity (see [25]) It is based on the classical Ricardian comparative advantage concept.

Balassa-Hoover Index (BHI) is the key index when we want to analyze revealed comparative advantages in a spatial setting. The formula of BHI is the following:

$$\mathrm{BHI}_{si} = \frac{\frac{x_{si}}{X_i}}{\frac{x_s}{X}}$$

where x_{si} is the number of employed people in region s and in industry i, (Xsi/Xi) is the corresponding share for all sub-regions. If $\mathrm{BHI}_{si} \geq 1$, there is a revealed comparative advantage in relation to all regions. In this way, we can understand whether regions have some comparative advantages which can provide a fundamental economic base for progressive industrial development.

Another very useful index is the Region's Relative Specialization Index (RRSI). We use RRSI as a measure of comparative specialization of sub-regions. RRSI can be calculated in the following way:

$$(\mathrm{RRSI})_s = \sqrt{(1-\mathrm{BHI}_{i1})^2 + (1-\mathrm{BHI}_{i2})^2 + \ldots + (1-\mathrm{BHI}_{\mathrm{in}})^2}$$

The higher the RRS index is the more specialized the structure of the manufacturing industry is in the region. If the structure of a region is homogenous, the RRSI obtains value zero. If the RRSI $\neq 0$, the industrial structure of a region differs from the country's average. The higher the RRSI value, the more a region is different from the whole country.

The idea is that the revealed comparative advantages provide 'objective' tools in which actual historical data for trade or employment reveals actual comparative advantages. If a region exports a given sectors' products in a quantity that outsizes both the market for those products and the general regional trade patterns, the region is presumed to have an advantage in this sector. Often in the S3 approach, micro-based qualitative indicators and a bottom-up approach help to make local tailoring possible by providing understandings of the specific characteristics of a given region. However, in order to compare regions and regions' relative differences for a given phenomenon of interest with 'objective' measuring tools, quantifiable macro-indicators are needed.

Two recent publications analyzing Smart Specialization in Finland show different methods of applying these quantifiable macro-indicators. Haukioja et al. [21] presents a Labour Intensity of Manufacturing Index (LIMI) measuring regions' manufacturing workforce's share of the total regional workforce in order to evaluate how dependent regions are on industrial manufacturing. Regions with a higher LIMI-share may have comparative advantages in manufacturing, but may also be more susceptible to risks in the case of rapid technological changes to manufacturing. Karppinen et al. [22] present analyses of Regions' Relative Specialisation Index (RRSI), based on a Balassa-Hoover (see [25]) index for the industry. This thorough analysis of Finland identifies no less than 432 relative advantage-subdivisions in the industrial sector alone [22]. This hints of many subnational comparative advantages around which the regions in question might build there Smart Specialization Strategy around.

For the long-term benefits of the comparative advantages, distinctions between various types of technologies and exports might be useful. Technologies and local business ecosystems that are simple to copy and easy to move to tend to be of reduced value and thus not a source of long-run rents [16]. More complex and difficult to imitate technologies are more sticky in space and holds greater promises for the regions and firms in which they are created. An alternative to revealed comparative advantages, that is to empirical evidence of actual trade-flows or employment patterns, are overall rankings of competitiveness. Global competitiveness rankings have been used to analyze countries' perspectives on Industry 4.0 [9]. In order for industrial manufacturing companies in a given country to thrive on open international markets, these are indeed relevant data. Fortunately, a well of global and refined data already exists in this regard. Competitiveness has been at the forefront of national economic policies around the world since the groundbreaking work of Michael Porter more than 25 years ago, and it has long forked from its firm-level origins into additional discussions of national competitiveness [26]. An important indicator of national competitiveness is the Global Competitiveness Reports of the World Economic Forum. Currently, these reports include five unique four analytical domains: (1) Enabling environment, (2) human capital, (3) markets and (4) innovation ecosystem. As the reader might note, this has certain overlaps with other elements of our model, i.e. they might be considered a reflection of the S3-triangle as a whole.

Resilience Base: Industrial and Business Diversity

Regional specialization strongly depends on the industrial structure actually present in a region, and as a result shows clear traits of embeddedness and path dependencies [11]. In many studies of economic resilience, much effort is attributed to the development of factors and measures representing economic and related resilience. In our S3 approach, we have focused on industrial and business diversity analysis. Resilience analyses are relevant and important because they help regions to adjust themselves to withstand and adjust to economic shocks. If the resilience level is very weak, even advanced industrial sectors can collapse and ruin all efforts to develop Industry 4.0 competences and capacities. We apply the Herfindahl-Hirchman Index (HHI) to the Finnish sub-region data. Our data includes 71 sub-regions and 24 industrial sectors [24, 21]. Our HHI analyses help decision-makers in the private and public sectors to identify resilience levels of the regions. The HHI formula is the following:

$$\mathrm{HHI}_s = \sum_{i=1}^{n} \left(\frac{x_1}{x} \right)_s^2$$

where x_i is the number of employed people in the industrial sector (i) and x is a total number of people employed in all industrial sectors in region (s) and n is a number of industrial sectors (n). HHI-index is calculated as the sum of squared industry shares for each sub-region.

This kind of basic resilience analysis informs decision-makers about industrial resilience levels which vary much in different regions and spatial communities. Resilience analyses help decision-makers to identify risks of industrial policy and understand strategic trade-offs between comparative advantage analyses and resilience analyses. Both "sides" of industrial policy are relevant for Industry 4.0 strategy in the national level.

Business Renewal, Innovation Ecosystem Dynamics

The fourth critical analysis tool of S3 is focused on the business renewal and innovation ecosystem. In general, business demographics size matters in Industry 4.0. Current Industry 4.0 research reveals that small- and medium-sized enterprises (SMEs) often have more challenges to adapt to Industry 4.0 approach [27, 28]. This is a good reason to pay more attention to spatial variation in entrepreneurship and business demographics.

The smart region is an innovative region that is flexible in adapting to economic shocks. Adaptability is measured by the CDI index. The CDI index is calculated for the whole regional enterprise stock:

$$\mathrm{CDI}_s = \left[\frac{EN_e + EN_e}{T_e} \right]_s$$

where EN_e = number of start-ups in the area, EX_e = number of deaths in the area and T_e is the number of enterprises in the area. The higher the CDI, the better the ability of the region to renew its business and industrial activity.

Entrepreneurial Discovery Process (EDP)

In Smart Specializations Strategies, it is presupposed that choices of technological domains reflect regions' economic, technological, and research specialization [11]. The principle of 'entrepreneurial discovery processes' (EDP) then ensures that the choice of specialization hinges on capabilities already existing in the region. There are very close links between this idea of EDP and spatial innovation ecosystems.

Innovation ecosystem is, therefore, a key component of the Smart Specialization Strategy. In Fig. 3, we report Big Data analysis of the global start-up ecosystem of manufacturing. This study aims to identify digital trends across industries and to map emerging sectors by using co-word and social network analysis. As the industrial landscape has become complex and dynamic due to the rapid pace of technological change and digital transformation, identifying industry trends and emerging business

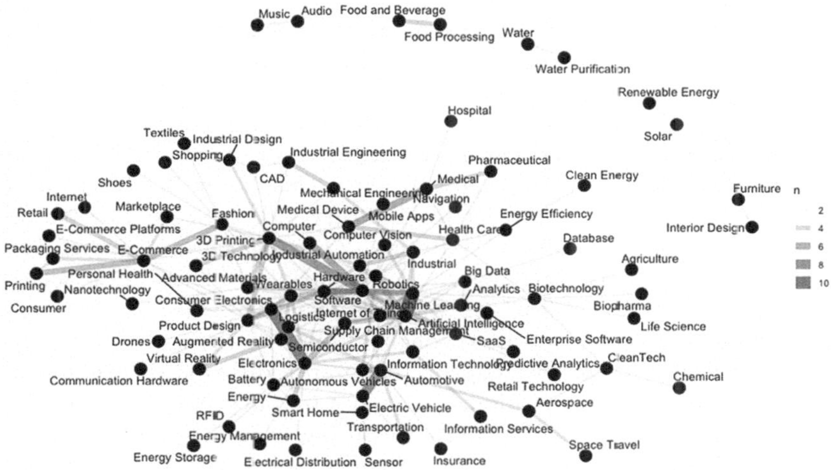

Fig. 3 Global start-up ecosystem of the Manufacturing cluster: A Big Data analysis [29]

areas can be critical for strategic planning and investment policy. This study examines industry and technology profiles of top start-ups across sectors and studies in which areas early-stage companies specialize. We have applied co-word analysis to reveal co-occurrences of keywords or key phrases related to technology and industry profiles of early-stage companies and then use social network analysis to visualize industry structure and to identify trends from word co-occurrence. The results obtained from the analysis show in which sub-industries digital technologies are penetrating and what new sectors are emerging [29].

We can see from the results (visible in Fig. 3) that

- The global start-up ecosystem of the manufacturing cluster is very diversified
- There strong links between robotics and machine learning, electronics and semiconductor (which is expected)
- There is a strong link between information tech, vehicles, and electric vehicles
- Industrial cluster centers are e-commerce, 3D printing, and energy utilities business
- Many elements of the global start-up ecosystem are isolated and not very networked.

The start-up ecosystem of Finland has been visualized and analyzed in a similar manner and the analysis shows not only results for Finland, but that it is possible to do both global analyses and analyses of smaller spatial entities, such as regions or countries. Among the 427 included start-ups in Finland, the role of manufacturing companies is very minor, with less than 3 percent of the total population situated in this sector. However, most start-ups are in sectors relevant for Industry 4.0, such as Information Technology (89 companies), Software (72 companies), Internet (45 Companies), Artificial Intelligence (43 companies). The largest sector outside this field is Health Care with 36 companies. We can note that Finnish start-up ecosystems need collaboration with the global start ecosystem to gain a competitive edge in the future. This kind of analysis can help decision-makers plan national Industry 4.0 strategy roadmaps. Similar analyses can be made for all countries with Big Data analytics. Again, we can note that many elements of the start-up ecosystem are quite isolated.

2 Technology Innovation, Markets and Business Modeling

Nations have the opportunity to pursue innovation and increase their technological competencies as the main method of supply-side preparation for potential technological shifts. It is possible to analyze preparedness based on this step through several different approaches. Common approaches are for example patent and trademark analyses [30] or bibliometric analyses. In the context of preparedness for increased automation, indicators of digital readiness may, however, be a useful tool allowing also for the comparison between different entities. With some modifications we can turn the European Union's Digital Economy and Society Index (DESI) [31] into a new Manufacturing 4.0-index in which the main dimensions are based on the characterization of the four main features of Industry 4.0: Interconnectivity, Data, Integration, and Innovation [32, 33]. In addition, we add the important fifth dimension of the availability of necessary human resources with the required digital skills.

This operationalization can be seen below in Table 1 which also reports the results for Finland and its ranking against the other EU27, Iceland, Norway, Montenegro, North Macedonia, Serbia, Turkey, and Bosnia and Herzegovina.

The analysis shows that Finland has one of the most conducive foundations for digital advancement in Europe, as it consistently ranks at or near the top across the five dimensions of digital Manufacturing 4.0-readiness.

For a region to succeed in a competitive international environment, it is not enough to have the right competences and enabling conditions. The industrial firms of the region must also function with business models fitting their respective markets. Filtering and prioritizing futureproof markets and business models are therefore key elements of the 'demand'-side of coping with technological changes. Two new and key methods of analyzing market change preparedness are the analysis of Long-Term Business Opportunities and Business Model-Based Filtering Analysis.

The analysis of Long-Term Business Opportunities is a means to explore baseline projections of the long-term business of e.g. manufacturing industries in a given country [modified from [34]). The baseline projections combine long-term projections of GDP growth for various countries [35], manufacturing outputs based on the World Input-Output Database (WIOD; [36]), national data on employment by manufacturing sectors, and calculations of Revealed Comparative Advantages. This

Table 1 Digital skills in Finland: Key results

	Variable	*Variable*	*Result*	*Rank*
Interconnectivity	Enterprises provide more than 20% of the employed persons with a portable device that allows a mobile connection to the Internet for business use	E_EMPMD2_ GT2020 2018	62%	2
	Share of turnover from web sales, business-to-business and business-to-government	ISOC_EC_ EVALN2	4%	8
Data	Using cloud computing services	ISOC_ CICCE_USE 2018	65%	1
	Enterprise analyzing big data from any data source	ISOC_EB_BD 2018	19%	5
Integration	Share of enterprises with ERP systems to share information internally	IE_ERP1 2017	39%	9
	Share of enterprises using CRM-systems	ISOC_ BDE15DEC 2011	39%	7
Innovation	Use 3D printing	ISOC_EB_ P3D 2018	7%	1
	Use robots	ISOC_EB_ P3D 2018	10%	2
	Cumulated Horizon 2020-funding for ICT-related projects	Data from DG Connect		7
Digital skills	Employed ICT specialists as share of total employment	ISOC_SKS_ ITSPT 2018	7.2%	1
	Percentage of individuals with basic or above basic overall digital skills	ISOC_SK_ DSKL_I 2017	76%	6
	STEM graduates at Master's or equivalent level per thousand inhabitants	EDUC_UOE_ GRAD04 2017	7.7%	4
Sum	Finland—Average ranking in Europe			4.4

framework provides a simple *ceteris paribus* analysis of whether existing trade patterns trend upwards or downwards. An analysis of the data for Finland shows that certain industrial sectors projects to increase their share of the Finnish economy (e.g. manufacture of machinery and equipment, and manufacture of paper and paper products), while growth potentials in other sectors lack behind (e.g., manufacture of food products, and manufacture of textiles).

Since the World Input-Output Database is so detailed, the information can also be used to forecast shifts in individual export markets. Below is an example again based on the latest available information for Finland (Table 2), which does not quite show radical shifts, but still hints of ongoing changes. China, perhaps not so surprisingly, looks to become a bigger and more important market for Finland than e.g. the United Kingdom and Russia, due to the projected larger growth in GDP. Similar tables can be made for all countries included in the WIOD-database and the OECD GDP projections. It is also possible to break the numbers down to subnational levels in order to analyze which regions, again ceteris paribus, which are better positioned for long-term business opportunities.

Table 2 Top 10 manufacturing sectors and markets (excluding Finland and the rest of the world)

	2012–2014	*2035*
1	Market: Germany *Paper and paper products*	Market: Germany *Paper and paper products*
2	Market: Sweden *Coke and refined petroleum*	Market: Sweden *Coke and refined petroleum*
3	Market: USA *Paper and paper products*	Market: USA *Paper and paper products*
4	Market: Germany *Basic metals*	Market: Sweden *Chemicals and chemical products*
5	Market: Sweden *Chemicals and chemical products*	Market: Germany *Basic metals*
6	Market: Germany *Printing and reproduction of recorded media*	Market: China *Paper and paper products*
7	Market: United Kingdom *Paper and paper products*	Market: Germany *Printing and reproduction of recorded media*
8	Market: Sweden *Basic metals*	Market: Sweden *Basic metals*
9	Market: USA *Printing and reproduction of recorded media*	Market: United Kingdom *Paper and paper products*
10	Market: Russia *Chemicals and chemical products*	Market: China *Machinery and equipment*

During the latest decades, much of business research has focused on Business Model Innovation [37]. In recent years, this has also led to an academic focus on business model innovation through Industry 4.0 (e.g. [38]). Beyond the Industry 4.0-literature, there are new and inspiring approaches to business modelling like "happiness based business models" [39], business models based on "platform thinking" [40] or "sustainable socially responsible and ethically oriented" business models [41]. It is good to remember that firms can select different approaches to their business model. In Table 3 we present key alternatives for export-oriented business models (Table 3).

Business models can also be simplified according to the main market targeted. Modern companies typically work in Business-to-Business (B2B) markets, Business-to-Consumer (B2C) markets, or Business-to-Government (B2G) markets. In the future, business models will also increasingly include Business-to-Digital Networks (B2DN) markets, as buying and selling will be done on account of algorithms in digital networks. Future B2DN-models can be related to all three other types of models today, as the digital models exceedingly spread.

Global markets can be filtered to give ideas about which business model that is must promising for given markets, as well as which markets

Table 3 Typical business models [42]

Business model	*Definition*
Direct Sales	The company itself sells directly to foreign end customers either in Finland or locally.
Resale	The company sells to resellers who sell directly or through intermediaries to the end customer.
Licensing	The technology (or equivalent) is made available to another company to package it into a product or service to be sold for a license fee.
Franchising	Foreign, local operator operates in accordance with the business concept developed in Finland.
Associated Company	A foreign owned company with minority share sells a product or service locally to foreign customers.
Joint Venture	Equally owned (50/50) foreign company that sells in the local market.
Subsidiary	Wholly owned (or majority owned). The parent company has a majority of the shares, participations or other voting rights in the subsidiary company. The parent company is required to prepare consolidated financial statement, which records the profit or loss generated by the foreign affiliate.

are most promising for a given business model. We can call this a Business Model-Based Filtering Analysis. In general, the filtering model is based on three main market criteria: The absolute size of the market, the size of the sectors per capita, and the sectors' share of the overall economy (data from the IMF [43]).

As we know, three strategic arenas of innovation are technological innovation, business model innovation and social innovation [44, 30]. Business model innovation has traditionally based on the identification of consumers and end-users in national and global markets. For Industry 4.0 strategy not only supply-side analyses (technology push) are relevant, but there is a need to think demand-side analyses (technology pull). Then business model innovation plays an important role. The scaling and scoping of Industry 4.0 technology innovations require new business model innovations, which are based on professional export and import strategies with data lake filtering, Big Data analytics and knowledge management. It is good to understand that there is not a linear path to the big success of Industry 4.0 strategy. Often even very successful companies are not always winning in foreign markers, but they can learn from international markets and their own business failures. It is good to remember that both technology push and technology pull factors of Industry 4.0/ Manufacturing 4.0 strategy need professional attention.

3 Technology Foresight

Technology foresight is closely linked to national innovation policy, technology policy, science policy, and education policy work. Between these policies, there are various strategic trade-offs and national planning needs. The typical argument of technology foresight has linked to supply-side (technology push) analyses, but not so much to the demand-side (technology pull). There are six key frameworks of foresight (cf. Table 4): (1) predictive frame, (2) planning frame, (3) scenaric frame, (4) visionary frame, (5) critical frame, and (6) transformative frame [45]. There can be both explorative and normative analyses in the field of technology foresight. The level of perceived unpredictability has impacts on a selected frame of technology foresight. There are many social and economic benefits of technology foresight.

These kinds of technology foresight aspects are relevant also in the Industry 4.0 strategy discussion, because many current Industry 4.0 analyses are more linked to supply-side (technology push) analyses. In the best

Table 4 Frameworks of foresight and benefits and inputs for Industry 4.0 and Manufacturing 4.0

Frameworks of foresight	*Benefits and inputs for Industry 4.0/Manufacturing 4.0 strategy*
Predictive frame	Trend predictions and baseline scenarios of Industry 4.0/ Manufacturing 4.0 developments
Planning frame	Industrial planning of IND4.0/MFG4.0
Scenaric frame	Alternative development strategies of IND4.0/MFG4.0, Risk analyses of IND4.0/MFG4.0
Visionary frame	Long-run visions of IND4.0/MFG4.0 developments
Critical frame	Risk (forecasting what if? -scenarios) and uncertainty (backcasting scenarios) analyses of IND4.0/MFG4.0 development
Transformative frame	Roadmaps of IND4.0/MFG4.0 development

case, technology foresight helps decision-makers to build strategic roadmaps and assess alternative technology choices. Typical foresight processes include (1) the use of foresight methods, (2) network and stakeholder analyses and (3) decision-support for decision-makers and the pragmatic use of decision models [46]. These three pillars of foresight are needed to provide "fully-fledged foresight" with diagnosis, prognosis and prescription phase of technology foresight. In this way, technology foresight must be linked to actual processes of networking and decision-making.

In the field of technology foresight, one relevant framework is Gartner Hype Curve analysis, which focuses on digital technologies. Gartner Hype Cycle methodology gives us a view of how a technology or application will evolve over time, providing a sound source of insight to manage its deployment within the context of various business fields. The Gartner Hype Cycle approach is providing an important perspective to technological transformation in the VUCA environment [1], which SMEs and corporations face in the global economy.

An Example of Global Technology Foresight: Technology Power Index Analysis of Digital ICT Technologies

Digital and disruptive technologies create the most economic growth and productivity. Here we analyze longitudinal data of the 2008–2017 Gartner Hype Cycles and key digital ICT technologies in the world. Gartner Hype Cycle analyses have a strong influence on large companies' technology strategies and investment decisions. In this sub-section, we present the key results of Technology Power Index Analysis, TPIA [47]. The TPIA is based on the ranking positions of technologies and the power index of each technology (151 technologies) in the yearly Gartner evaluations from the year 2008 to 2017 [48]. The technologies ranked first to receive the highest power index numbers, and the technologies which are ranked lowest in the Gartner Hype Cycle have the smallest index numbers. Based on these TPIA calculations, all technologies receive a TPIA sum of technology power numbers. The larger the sum each technology gets, the more powerful the analyzed technology is. According to our analyses, the top 10 most powerful technologies are Surface Computers, Consumer Telematics, Mobile OTA Payment, Location Intelligence, Enterprise 3D Printing, Consumerization, Biometric Authentication Methods, Text Analytics, People-Literate Technology, Neurometric Hardware and In-Memory Analytics.

The lowest TPIA index technologies are held by new technologies in the "Technology Trigger Phase" of the Gartner Hype Curve. None of these low TPIA index-value technologies and innovations are fully ready to be used. It will take from five to ten years for them to become more powerful. The most recent Gartner reports for 2018–2020 provide some new insight into global technology foresight. New technology challenges are Augmented analytics, Quantum computing and Autonomous Things and Artificial Intelligence (AI). *Augmented analytics* reflects a third major wave for data and analytics capabilities. Data scientists are able to use automated algorithms to explore more hypotheses [49]. *Quantum computing* is based on the quantum state of subatomic particles. Quantum computers are an exponentially scalable and highly parallel computing model [49]. Autonomous things mean Robotics, Vehicles, Drones, Appliances and Agents. They can operate in four environments: sea, land, air and digital. Autonomous things use Artificial Intelligence (AI) to perform tasks traditionally done by humans [49].

The analysis of the Gartner Hype Curve helps leaders to understand the dynamics of on-going technological disruption, which is extremely important for SMEs and corporate leaders in being able to foresight the future of digital ICT technologies. Technological transformation, in particular, is changing many basic assumptions of business management and strategic planning. The digital transformation process can be estimated from the yearly results of Gartner's Hype Curve analysis. This technology foresight study reveals that companies have to take significant risks when making technological choices. It can be argued that technology risks can be managed, but not completely eliminated by technology foresight. By using futures studies and technology foresight methods it is possible to engage relevant stakeholders in the region in ideations on how these technologies might affect business in the region, as well as how the region could move forward in order to generate opportunities based on the emerging technologies.

4 Conclusions

This chapter highlights that in the discussions about Industry 4.0 and Manufacturing 4.0 development demand and technology pull factors are often forgotten and the technology push approach dominates discussions. We should understand also the demand side of the Industry 4.0 orchestration in a connection with consideration of the development of Industry 4.0 know-how and competences. The analytical thinking behind the Industry 4.0 approach is not just a question of supply-side and the production modernization and fast digitization of super-connected ubiquitous production. Supply chain management is an elementary part of Industry 4.0 expertise and development. In practice, this means that the supply and demand side (eCommerce and consumption on the domestic and international markets) analyses must be linked together. This means also a new kind of orchestration challenge throughout the whole supply chain and a new kind of innovative business model development.

In this chapter, we have addressed the general challenges of anticipation and the development of Industry 4.0 and Manufacturing 4.0 strategy. Foresight analyses can provide an overview of how the transition from Industry 3.0 to Industry 4.0 is taking place. Generally speaking, the transition from Stage 1.0 and Stage 2.0 to Industry 4.0 is extremely challenging for any operator or company. Conversely, the transition from Industry 3.0 is, of course, one step easier, as the industrial companies that have reached this Industry 3.0 stage, already have proven capabilities and intellectual capital to support the transition to Industry 4.0 level. Generally speaking, we can say that the proven abilities in international competition are helping companies move into Industry 4.0 level. Digital learning and learning, in general, is an important part of the transition to Industry 4.0. Manufacturing 4.0 and Industry 4.0 target is already challenging because of the need to combine cognitive ergonomics with physical ergonomics. This requires in-house testing, innovation and experimentation, for which smaller companies, in particular, have fewer resources than large corporations. It is evident that larger companies act as drivers of change in many modern industries, shaping their industry cultures and practices through their own policies and standards. In particular, this affects the operational supply-side logistics and supply chains. On the other hand, small businesses firms tend to be innovative and agile, and through their own start-up ecosystem, they can develop into major players in the Industry 4.0, too.

Industry 4.0 is a challenge for both small and large companies, and often local ecosystems of innovation play a major role in regional industrial changes and transformations. As a key result, in this chapter, we have presented the European approach of the Smart Specialization Strategy (S3). We argue that achieving Industry 4.0 level in the industry will be easier if and when decision-makers have knowledge of the state of smart specialization regionally. Of course, factors such as population, employment and migration are important background factors, but the revealed comparative advantages, resilience levels, renewal and creativity processes, and entrepreneurial culture are the fundamental pillars on which the transition to Industry 4.0 is built regionally. When reliable information is available about these strategic key factors, making future investments is also easier.

Technology Foresight offers its own perspectives on Industry 4.0 know-how and development. Foresight can provide six different perspectives on Industry 4.0 development: (1) Predictive frame, (2) Planning frame, (3) Scenaric frame, (4) Visionary frame, Critical frame, and (6) Transformative frame. The Predictive frame can be used to produce statistical forecasting analyses of socio-economic and technical forecasts and baseline scenario development for Industry 4.0 developments. The Planning frame can be used to generate operational and strategic plans for the development of Industry 4.0 supply chains. The Scenaric frame can be used to generate forecasting and backcasting scenarios for Industry 4.0 and Manufacturing 4.0 developments. Through the Visionary frame, visions of Industry 4.0 developments can be created and produced. Through the Transformative frame, it is possible to produce technology roadmaps for Industry 4.0 development. There are many possibilities to apply foresight tools and methods to develop Industry 4.0 and Manufacturing 4.0 know-how and competences.

The Garner Hype Curve Technology Foresight presented in this article is a good example of a foresight tool that can be used to evaluate the evolution of very large ICT and digital technologies over time. Through our

TPIA technology foresight analysis, we can create a better understanding of digital technology development, helping companies to evaluate their own technology choices and the risks involved in technology choices.

References

1. Kaivo-oja, Jari & Lauraeus, Theresa (2018). "The VUCA approach as a solution concept to corporate foresight challenges and global technological disruption", *Foresight*, 20(1): 27–49.
2. Suni, Paavo & Vihreälä, Vesa (2016). "Finland and Its Northern Peers in the Great Recession". ETLA Reports No. 49.
3. Heikkilä, Jussi, Martinsuo, Miia & Nenonen, Sanna (2018). "Backshoring of production in the context of a small and open Nordic economy". *Journal of Manufacturing Technology Management*, 29(4): 658–675.
4. Muscio, Allesandro & Ciffolilli, Andrea (2020). "What drives the capacity to integrate Industry 4.0 technologies? Evidence from European R&D projects". *Economics of Innovation and New Technology*, 29(2): 169–183.
5. Ciffolilli, Andrea & Muscio, Alessandro (2018). "Industry 4.0: National and regional comparative advantages in key enabling technologies". *European Planning Studies*, 26(12): 2323–2343.
6. Schumacher, Andreas, Erol, Selim & Sihn, Wilfried (2016). "A maturity model for assessing Industry 4.0 readiness and maturity of manufacturing enterprises". *Procedia CIRP*, 52: 161–166.
7. Akdil, Kartal Y., Upstundag, Alp & Cevikcan, Emre (2018). "Maturity and Readiness Model for Industry 4.0 Strategy". In *Industry 4.0: Managing The Digital Transformation*. Springer Series in Advanced Manufacturing, Springer: Cham.
8. Lobova, Svetlana V., Bykovskaya, Natalia V., Vlasova, Irina M., Sidorenko Olga, V. (2019). "Successful Experience of Formation of Industry 4.0 in Various Countries." In: Popkova, E., Ragulina, Y.V. & Bogoviz, A. V. (eds.) *Industry 4.0: Industrial Revolution of the 21st Century*. Studies in Systems, Decision and Control, Cham: Springer.
9. Sung, Tae Kyung (2018). "Industry 4.0: A Korea Perspective". *Technological Forecasting & Social Change,* 132: 40–45.
10. Cretu, Corina (2016). "Preface". In Gianelle, Carlo, Kyriakou, Dimitris, Cohen, Caroline & Przeor, Marek (2016). Implementing Smart Specialisation Strategies: A Handbook. Brussels: European Commission.
11. D'Adda, Diego, Guzzini, Enrico, Iacobucci, Donato & Palloni, Roberto (2019). "Is Smart Specialisation Strategy coherent with regional innovative capabilities?". *Regional Studies,* 53(7): 1004–1016.

12. Foray, Dominique (2015). "Smart Specialisation: Opportunities and Challenges for Regional Innovation Policy." London: Routledge.
13. Bzhalava, Levan, Kaivo-oja, Jari, Hassan, Sohaib H. (2018). "Data-based startup profile analysis in the European Smart Specialization Strategy: A text mining approach". *European Integration Studies*, 12(1): 118–128.
14. Bailey, David & De Propris, Lisa (2019). "Industry 4.0, Regional Disparaties and Transformative Industrial Policy". *Regional Studies Policy Impact Books,* 1(2): 67–78.
15. Foray, Dominique (2012). "Economic fundamentals of smart specialisation". *Economies*, 83(2): 55–78.
16. Balland, Pierre-Alexandre, Boschma, Ron, Crespo, Joan & Rigby, David L. (2019). "Smart specialization policy in the European Union: Relatedness, knowledge complexity and regional diversification." *Regional Studies*, 53(9): 1252–1268.
17. Foray, Dominique (2016). "On the policy of smart specialization strategies". *European Planning Studies*, 24(8): 1428–1437.
18. Coffano, Monica & Foray, Dominique (2014). "The centrality of entrepreneurial discovery in building and implementing a smart specialisation strategy". *Scienze Regionali*, 13(1): 33–50.
19. Lafuente, Esteban, Vaillaint, Yancy & Vendrell-Herrero, F. (2017). "Territorial servitization: Exploring the virtuous circle connecting knowledge-intensive services and new manufacturing businesses". *International Journal of Production Economics*, 192: 19–28.
20. Bellandi, Marco, De Propris, Lisa & Santini, Enrica (2019). "Industry 4.0+ challenges to local productive systems and place based integrated industrial policies.". In Bianchi, Patrizia, Duran, Clemente R. & Labory, Sandrine (eds.). *Transforming Industrial Policy for the Digital Age: Production, Territories and Structural Change*. Edward Elgar Publishing Limited, Cheltenham, UK.
21. Haukioja, Teemu, Kaivo-oja, Jari, Karppinen, Ari & Vähäsantanen, Saku (2018). "Identification of smart regions with resilience, specialisation and labour intensity in a globally competitive sector – Examination of LAU-1 regions in Finland". *European Integration Studies*, 12(1): 50–62.
22. Karppinen, Ari, Aho, Samuli, Haukioja, Teemu, Kaivo-oja, Jari & Vähäsantanen, Saku (2019). "Alueiden Älykäs Erikoistuminen Suomessa: Aluekehittämisen Indikaattorianalyysi." Tulevaisuuden Tutkimuskeskus, TUTU eJULKAISUJA 4/2019.
23. Paliokaitė, Agne, Martinaitis, Zilvinas, Reimeris, Ramojus (2015). "Foresight methods for smart specialisation strategy development in Lithuania". *Technological Forecasting and Social Change*, 101(1): 185–199.
24. Kaivo-oja, Jari, Vähäsantanen, Saku, Karppinen, Ari & Haukioja, Teemu (2017). "Smart specialization strategy and its operationalization in the regional policy: case Finland". *Business, Management and Education*, 15(1): 28–41.

25. Balassa, Bela & Noland, Marcus (1989). “Revealed Comparative Advantage in Japan and the United States”. *Journal of International Economic Integration.* 4 (2): 8–22.
26. Aiginger, Karl & Firgo, Matthias (2017). “Regional Competitiveness: Connecting and old concept with new goals.” In Huggins, Robert & Thompson, Piers (eds). *Handbook of Regions and Competitiveness: Contemporary Theories and Perspectives on Economic Development.* Edward Elgar Publishing Limited: Cheltenham.
27. Sommer, Lutz (2015). “Industrial revolution – industry 4.0: Are German manufacturing SMEs the first victims of the revolution?”. *Journal of Industrial Engineering and Management,* 8(5): 1512–1532.
28. Müller, Julian M., Buliga, Oana & Voigt, Kai-Ingo (2018). ”Fortune favors the prepared: How SMEs approach business model innovations in Industry 4.0”. *Technological Forecasting and Social Change,* 132: 2–17.
29. Bzhalava, Levan, Hassan, Sohaib S., Kaivo-oja, Jari & Olsson, Bengt K. (forthcoming). “Mapping the wave of industry digitalization by co-word analysis”. Forthcoming.
30. Kaivo-oja, Jari (2016). “Benchmarking Analysis of Patent and Trademark Applications in the European Union: Comprehensive Innovation Policy Evaluation for Years 1960–2013”. *European Integration Studies,* 10(1): 169–190.
31. European Commission (2019). “Digital Agenda key indicators.” Accessed 31.01.2020 at https://digital-agenda-data.eu/datasets/digital_agenda_scoreboard_key_indicators/indicators.
32. Cheng, Guo-Jian, Liu, Li-Ting, Qiang, Xin-Jian & Liu, Ye (2016). “Industry 4.0 Development and Application of Intelligent Manufacturing. In: International Conference on Information Systems and Artificial Intelligence (ISAI). IEEE Xplore.
33. Knudsen, Mikkel S., Kaivo-oja, Jari & Lauraeus, Theresa (2019). ”Enabling Technologies of Industry 4.0 and Their Global Forerunners: An Empirical Study of the Web of Science Database”. In: Uden, L., Ting, I.H. & Corchado, J. (eds). *Knowledge Management in Organizations. KMO 2019. Communications in Computer and Information Science,* Vol 1027: 3–13. Cham: Springer.
34. Small Great Nation (2018). ”Small Great Nation: Challenges and Opportunities”. Accessed 31.1.2020 at https://sgnation.dk/application/files/2715/1548/7498/Summary_Small_Great_Nation_-_challenges_and_opportunities.pdf
35. OECD (2018). “Long-term baseline projections, No. 103”, *OECD Economic Outlook: Statistics and Projections.* (database).
36. Timmer, Marcel P., Dietzenbacher, Erik, Los, Bart., Stehrer, Robert & de Vries, Gaaitzen J. (2015). ”An Illustrated User Guide to the World Input-Output Database: the Case of Global Automotive Production”. *Review of International Economics,* 23: 575–605.

37. Chesbrough, Henry (2010). "Business Model Innovation: Opportunities and Barriers". *Long Range Planning*, 43(2–3): 354–363.
38. Ibarra, Dorleta, Ganzarain, Jaione & Igartua, Juan I. (2018). "Business model innovation through Industry 4.0: A review". *Procedia Manufacturing*, 22: 4–10.
39. Fagerström, Arne & Cunningham, Gary (eds.) (2017). "A Good Life for All: Essays on Sustainability Celebrating 60 Years of Making Life Better". Mjölby: Atremi AB.
40. Hakanen, Esko (2018). "Platform-based Exchange: New Business Models in Technology Industries. Aalto University Publication Series, Doctoral dissertations 250/2018. Helsinki: Aalto University.
41. Geissdoerfer, Martin, Vladimirova, Doroteya & Evans, Steve (2018). "Sustainable business model innovation: A review". *Journal of Cleaner Production*, 198(10): 401–416.
42. Väisänen, Kim (2018). "Väärää vientiä. Mene itään tai länteen, mutta tee kotiläksysi". Helsinki: Alma Talent.
43. International Monetary Fund (IMF) (2019). "World Economic Outlook Databases and IMF Data Mapper". Web: https://www.imf.org/en/Publications/SPROLLS/world-economic-outlook-databases#sort=%40imfdate%20descending and https://www.imf.org/external/datamapper/NGDP_RPCH@WEO/OEMDC/ADVEC/WEOWORLD
44. Kaivo-oja, Jari (2011). "Futures of Innovation Systems and Systemic Innovation Systems: Towards Better Innovation Quality with New Innovation Management Tools". FFRC eBOOK 8/2011. Finland Futures Research Centre. Turku: University of Turku.
45. Minkkinen, Matti, Aufferman, Burkhard & Ahokas, Ira (2020). "Six foresight frames: Classifying policy foresight processes in foresight systems according to perceived unpredictability and pursued change". *Technological Forecasting and Social Change*, 149: 1–13.
46. Olsmats, Carl & Kaivo-oja, Jari (2014). "European packaging industry foresight study – Identifying global drivers and driven packaging industry implications of the global megatrends". *European Journal of Futures Research*, 2(1): 1–10.
47. Kaivo-oja, Jari, Lauraeus, Theresa & Knudsen, Mikkel S. (forthcoming). "Picking the ICT technology winners – longitudinal analysis of 21st-century technologies on the basis of the Gartner hype cycle 2008–2017: trends, tendencies and weak signals". *International Journal of Web Engineering and Technology*, accepted, forthcoming.
48. Gartner (2008–2017). "Gartner reports". Gartner Ltd.
49. Gartner (2019). "5 Trends Appear on the Gartner Hype Cycle for Emerging Technologies, 2019". Accessed 31.01.2020 at https://www.gartner.com/smarterwithgartner/5-trends-appear-on-the-gartner-hype-cycle-for-emerging-technologies-2019/.

Closing Words: Quo vadis Manufacturing 4.0

Mikael Collan and Karl-Erik Michelsen

How should we understand and explain the fundamental changes that take place in the manufacturing industry, when the ways of manufacturing change—advanced robots and other types of automation replace human workers, and advanced analytics are used to optimize and control what autonomous machines are doing? Are we in the early phase of the transformation or have we already passed the turning point where there is no turning back? These questions are troubling scholars, policy makers, and business managers. Nobody seems to have a clear vision of future and the confusion affects industrial, technology, economic, and social policy-making not only in Finland, but also in Europe and beyond.

This book takes a multidisciplinary approach to the complex issue that is Manufacturing 4.0—multidisciplinarity is something that the "original Industry 4.0" vision was full of, we must remember that it was a holistic vision of where the society is going as a whole. It is not an easy task to combine technical, economic and social aspects of Manufacturing in a scientifically credible narrative—there are too many contradictions and uncertainties. There are also too many positive expectations and promises

M. Collan (✉) • K.-E. Michelsen
LUT University, Lappeenranta, Finland
e-mail: mikael.collan@lut.fi; karl-erik.michelsen@lut.fi

M. Collan, K.-E. Michelsen (eds.), *Technical, Economic and Societal Effects of Manufacturing 4.0*,
https://doi.org/10.1007/978-3-030-46103-4_13

of technological advances and still relatively few concrete solutions to point at. And yet, nobody can deny the ongoing technological revolution that is changing the industrial and economic landscapes. Modern industrial and post-industrial societies are affected by radical new technologies even before they are fully implemented to and adopted by the industry.

This book reveals at least four important messages. First, Manufacturing 4.0 is today a predominantly technological issue and in that it is not one-sided—both the methods and materials of manufacturing are experiencing profound changes with additive manufacturing changing the way structures can be constructed, and new advanced materials are becoming available to be exploited in the future. When automation of industry, advanced robotics, and digitalization of manufacturing information are added to this mix, what can be seen is a complex multi-level transformation, a revolution, in how industrial systems are changing.

Second, the changes happening in the technologies propel changes in the business models that companies use—these in turn cause a shift in the architecture of businesses in the field of manufacturing to become more networked in those niches, where the networked architecture is more competitive than the traditional "factory". We are still taking the initial steps of this revolution from the point of view of business architectures, but history tells us that evolution of how business is being conducted carries a strong resemblance to natural evolution—the strong and the competitive will survive, while the weak perish. When business architecture needs start forcing changes in the development of technology and new technologies in manufacturing that carry great business-power emerge, then the speed of change will most likely pick up.

Third, Manufacturing 4.0 is also a social "program" that must try to tame the otherwise chaotic future that will be caused by the technological change. The Fourth Industrial Revolution continues on the evolution path that started more than two centuries ago. Industrial societies have adapted to the radical technological changes and created social and economic structures, which are flexible and resilient. Radical technologies of Manufacturing 4.0 challenge the current social and economic structures and promise futures, which are more efficient, more productive, and perhaps also more sustainable. Technology changes society, but technology is simultaneously changed by changes in the society. As this book points out, a likely outcome of this interaction is not a purely technologically deterministic future, but a "new society", which still carries forward sturdy structures from the past industrial revolutions, but also exhibits new structures, which force societies to make difficult policy decisions.

Fourth, Manufacturing 4.0 is a political program that tries to derail the current industrial path that for the past decades has concentrated on the manufacturing industries in the low-cost developing countries. Manufacturing 4.0 builds on an age-old scientific and technological tradition, which is the cornerstone of Western culture and seems to promise to bring the manufacturing industry back to the developing world in a new transformed form and to anew create a competitive edge for industry-based business, in what we now think of as post-industrial societies. In order for manufacturing to make a come-back to the Western world the society must provide it "food and nourishment" in terms of not only "allowing" this change to happen, but also by way of supporting it by providing a workforce with a relevant know-how and by adjusting how the society works.

It remains to be seen how massive the actual societal changes driven by the Industry 4.0 and Manufacturing 4.0 are—most likely we can understand it only after decades from today. Whether the fabric of the societies is flexible and able to withstand the changes, or whether the fabric will break and cause chaos is also an open question. What remains to be said here is that the only sure constant is change itself.

Index

M. Collan, K.-E. Michelsen (eds.), *Technical, Economic and Societal Effects of Manufacturing 4.0*,
https://doi.org/10.1007/978-3-030-46103-4

The manufacturer's authorised representative in the EU is Springer Nature Customer Service Centre GmbH, Europaplatz 3, 69115 Heidelberg, Germany. If you have any concerns regarding our products, please contact ProductSafety@springernature.com

Printed and bound by CPI Group (UK) Ltd, Croydon, CR0 4YY
30/09/2026
02225645-0006